Lecture Notes in Computer Science   11605

More information about this series at http://www.springer.com/series/7410

Jianying Zhou · Robert Deng ·
Zhou Li · Suryadipta Majumdar ·
Weizhi Meng · Lingyu Wang ·
Kehuan Zhang (Eds.)

Applied Cryptography and Network Security Workshops

ACNS 2019 Satellite Workshops, SiMLA, Cloud S&P, AIBlock, and AIoTS
Bogota, Colombia, June 5–7, 2019
Proceedings

 Springer

Editors
Jianying Zhou (ID)
Singapore University of Technology
and Design
Singapore, Singapore

Zhou Li
University of California
Irvine, USA

Weizhi Meng (ID)
Technical University of Denmark
Kongens Lyngby, Denmark

Kehuan Zhang
Chinese University of Hong Kong
Shatin, Hong Kong

Robert Deng
Singapore Management University
Singapore, Singapore

Suryadipta Majumdar (ID)
Massry Center for Business
University at Albany-SUNY
Albany, NY, USA

Lingyu Wang (ID)
Concordia University
Montreal, QC, Canada

ISSN 0302-9743   ISSN 1611-3349   (electronic)
Lecture Notes in Computer Science
ISBN 978-3-030-29728-2   ISBN 978-3-030-29729-9   (eBook)
https://doi.org/10.1007/978-3-030-29729-9

LNCS Sublibrary: SL4 – Security and Cryptology

This Springer imprint is published by the registered company Springer Nature Switzerland AG
The registered company address is: Gewerbestrasse 11, 6330 Cham, Switzerland

Preface

This post-proceedings contains the papers selected for presentation at ACNS 2019 satellite workshops, which were held in conjunction with the main conference (the 17th International Conference on Applied Cryptography and Network Security) during June 5–7, 2019, in Bogota, Colombia. The local organization was in the capable hands of Professors Valerie Gauthier-Umana from Universidad del Rosario, Colombia, and Martin Ochoa, Universidad del Rosario and Cyxtera Technologies; we are deeply indebted to them for their generous support and leadership to ensure the success of the event.

It was a new initiative for ACNS to set up satellite workshops in 2019. Each workshop provided a forum to address a specific topic at the forefront of cybersecurity research. In response to this year's call for workshop proposals, the following five workshops were launched.

- AIBlock – First International Workshop on Application Intelligence and Block-chain Security
- AIoTS – First International Workshop on Artificial Intelligence and Industrial Internet-of-Things Security
- Cloud S&P – First International Workshop on Cloud Security and Privacy
- PriDA – First International Workshop on Privacy-preserving Distributed Data Analysis
- SiMLA – First International Workshop on Security in Machine Learning and its Applications

This year, we received a total of 30 submissions. Each workshop had its own Program Committee (PC) in charge of the review process. These papers were evaluated on the basis of their significance, novelty, and technical quality. The review process was double-blind. In the end, 10 papers were selected for presentation at four workshops (PriDA was canceled), with an acceptance rate of 33%. The following paper received the ACNS 2019 Best Workshop Paper Award.

- "Risk-based Static Authentication in Web Applications with Behavioural Biometrics and Session Context Analytics," by Jesus Solano, Luis Camacho, Alejandro Correa, Claudio Deiro, Javier Vargas, and Martin Ochoa.

A couple of papers from the post-proceedings will be nominated for submission of an extended version to a special issue in the *International Journal of Information Security* published by Springer.

ACNS 2019 workshops were made possible by the joint efforts of many individuals and organizations. We appreciate Springer's strong support on our new initiative. We sincerely thank the authors of all submissions. We are grateful to the program chairs and PC members of each workshop for their great effort in providing professional reviews and interesting feedback to authors in a tight time schedule. We thank all the external reviewers for assisting the PC in their particular areas of expertise. We also thank the organizing team members of the main conference as well as each workshop for their help in various aspects.

Last but not least, we thank everyone else, speakers and session chairs, for their contribution to the program of ACNS 2019 workshops.

We believe this is a good start for the success of ACNS satellite workshops. We hope the existing workshops will keep growing and new workshops on emerging topics will be launched in the coming years. We expect it could provide a stimulating platform to discuss open problems at the forefront of cybersecurity research.

July 2019   Jianying Zhou
                                                    ACNS 2019 Workshop Chair

SiMLA 2019

First International Workshop on Security in Machine Learning and its Applications

Bogota, Colombia
June 5, 2019

Program Chairs

Zhou Li — University of California Irvine, USA
Kehuan Zhang — The Chinese University of Hong Kong, SAR China

Program Committee

Ninghui Li — Purdue University, USA
Yinqian Zhang — Ohio State University, USA
Di Tang — The Chinese University of Hong Kong, SAR China
Xiaokui Shu — IBM, USA
Jialong Zhang — ByteDance AI Lab, USA
Mohammad Al Faruque — University of California Irvine, USA
Xiapu Daniel Luo — The Hong Kong Polytechnic University, SAR China
Kai Chen — Institute of Information Engineering, Chinese Academy of Science, China
Zhe Zhou — Fudan University, China

Additional Reviewers

Guo, Zhixiu
Lu, Haochuan
Tan, Mingtian
Yuan, Xuejing

CLOUD S&P 2019

First International Workshop on Cloud Security and Privacy

Bogota, Colombia
June 5, 2019

Program Chairs

Suryadipta Majumdar	University at Albany, USA
Lingyu Wang	Concordia University, Canada

Web Chair

Sudershan L T	Concordia University, Canada

Program Committee

Vijay Atluri	Rutgers University, USA
Mauro Conti	University of Padua, Italy
Sabrina De Capitani di Vimercati	Universitá degli studi di Milano, Italy
Sara Foresti	Universitá degli Studi di Milano, Italy
Xinwen Fu	University of Central Florida, USA
Yuan Hong	Illinois Institute of Technology, USA
Patrick Hung	University of Ontario Institute of Technology, Canada
Yosr Jarraya	Ericsson Security, Canada
Ram Krishnan	University of Texas at San Antonio, USA
Zheli Liu	Nankai University, China
Rongxing Lu	University of New Brunswick, Canada
Ali Miri	Ryerson University, Canada
Makan Pourzandi	Ericsson Security, Canada
Indrakshi Ray	Colorado State University, USA
Pierangela Samarati	Universitá degli studi di Milano, Italy
Vijay Varadharajan	University of Newcastle, Australia
Mohammad Zulkernine	Queen's University, Canada

Additional Reviewers

Karmakar, Kallol Krishna
Shojafar, Mohammad
Zhang, Mengyuan

AIBlock 2019

First International Workshop on Application Intelligence and Blockchain Security

Bogota, Colombia
June 5, 2019

Program Chairs

Weizhi Meng   Technical University of Denmark, Denmark
Zhiqiang Liu   Shanghai Jiao Tong University, China
Chunhua Su   University of Aizu, Japan

Program Committee

Ashiq Anjum   University of Derby, UK
Man Ho Au   The Hong Kong Polytechnic University, SAR China
Raja Naeem Akram   Royal Holloway, University of London, UK
David Chadwick   University of Kent, UK
Konstantinos Chalkias   R3, UK and USA
Taolue Chen   Oxford University, UK
Dieter Gollmann   Hamburg University of Technology, Germany
Debiao He   Wuhan University, China
Georgios Kambourakis   University of the Aegean, Greece
Hyoungshick Kim   Sungkyunkwan University, South Korea
Hiroaki Kikuchi   Meiji University, Japan
Peter Lewis   Aston University, UK
Jiqiang Lu   Beihang University, China
Xiapu Luo   The Hong Kong Polytechnic University, SAR China
Felix Gomez Marmol   University of Murcia, Spain
Petr Novotny   IBM, USA
Jun Shao   Zhejiang Gongshang University, China
Seungwon Shin   KAIST, South Korea
Paolo Tasca   University College London, UK
Andreas Veneris   University of Toronto, Canada
Qianhong Wu   Beihang University, China

Additional Reviewers

Dai, Xiaopeng
Recchia, Alessandro
Wu, Lei

AIoTS 2019

First International Workshop on Artificial Intelligence and Industrial Internet-of-Things Security

Bagota, Colombia
June 5, 2019

Program Chairs

Robert Deng   Singapore Management University, Singapore
Sandra Rueda   Universidad de los Andes, Colombia

Organizing Chairs

Sridhar Adepu   SUTD, Singapore
John Henry Castellanos   SUTD, Singapore
Chuadhry Mujeeb Ahmed   SUTD, Singapore

Publicity Chair

Robert Kooij   SUTD, Singapore

Program Committee

Alvaro Cardenas   University of California Santa Cruz, USA
Ee-Chien Chang   National University of Singapore, Singapore
Debin Gao   Singapore Management University, Singapore
Luis Garcia   University of California Los Angeles, USA
Dieter Gollmann   TU Hamburg, Germany
Pieter Hartel   TU Delft, The Netherlands
Eunsuk Kang   Carnegie Mellon University, USA
Junyu Lai   UESTC, China
Elena Lisova   Malardalen University, Sweden
Javier Lopez   University of Malaga, Spain
Di Ma   University of Michigan, USA
Chris Poskitt   SUTD, Singapore
Pandu Rangan   IIT Madras, India
Rajib Ranjan Maiti   BITS-Hyderabad, India
Giedre Sabaliauskaite   SUTD, Singapore

Phani Vadrevu   University of New Orleans, USA
Yousheng Zhou   CQUPT, China

Additional Reviewers

Adepu, Sridhar
Ahmed, Chuadhry Mujeeb

Contents

SiMLA - Security in Machine Learning and its Applications

Risk-Based Static Authentication in Web Applications with Behavioral Biometrics and Session Context Analytics

Jesus Solano$^{(\boxtimes)}$, Luis Camacho, Alejandro Correa, Claudio Deiro,
Javier Vargas, and Martín Ochoa

Cyxtera Technologies, Coral Gables, USA
{jesus.solano,luis.camacho,alejandro.correa,claudio.deiro,
javier.vargas,martin.ochoa}@cyxtera.com

Abstract. In order to improve the security of password-based authentication in web applications, it is a common industry practice to profile users based on their sessions context, such as IP ranges and Browser type. On the other hand, behavioral dynamics such as mouse and keyword features have been proposed in order to improve authentication, but have been shown most effective only in continuous authentication scenarios. In this paper we propose to combine both fingerprinting and behavioral dynamics (for mouse and keyboard) in order to increase security of login mechanisms. We do this by using machine learning techniques that aim at high accuracy, and only occasionally raise alarms for manual inspection. Our combined approach achieves an AUC of 0.957. We discuss the practicality of our approach in industrial contexts.

Keywords: Behavioral dynamics · Static authentication ·
Machine learning

1 Introduction

With the increasing popularity of web services and cloud-based applications, we have also seen an increase on attacks to those platforms in the past decade. Several of those publicly known attacks have involved stealing of authentication credential to services (see for instance [10]). In addition to this, passwords are often the target of Malware (for instance banking related Malware such as Zeus [6] and its variants). So even if one would assume users are forced to select strong passwords (from the point of view of difficulty to guess), several studies have pointed out the challenges that password-based authentication pose for robust security [2,11].

In order to mitigate the risk posed by attackers impersonating legitimate users by means of compromised or guessed credentials, many applications use mechanisms to detect anomalies by analyzing the connection features such as incoming IP, browser and OS type as read by HTTP headers, among others.

© Springer Nature Switzerland AG 2019
J. Zhou et al. (Eds.): ACNS 2019 Workshops, LNCS 11605, pp. 3–23, 2019.
https://doi.org/10.1007/978-3-030-29729-9_1

Some of this context-based features have been also been discussed in the scientific literature [1]. However, there are some limitations of those defensive mechanisms, for example, if the anomaly detection is too strict, there could be false positives that would harm user experience and thus hurt the webservices from a business perspective. On the other hand, if a careful attacker manages to bypass such context-related filters, for instance by manipulating HTTP parameters, using VPN services, or ultimately using a victim's machine [12], then such countermeasures fall short to provide better security.

Behavioral biometrics [17] have been proposed in the literature as a strategy to enhance the security of both web and desktop applications. They have shown to work with reasonable accuracy in the context of continuous authentication [9, 19], when the monitoring time of mouse and/or keyboard activity is long enough. In the context of static authentication, where interaction during log-in time with users is limited, such methods are less accurate and may be impractical [13], unless long static authentication interactions are assumed. However, in today's internet of services, many websites rely on third parties for security related functionality, that is integrated in the form of external javascript snippets. In domains handling highly sensitive data such as banking, those services are often only allowed to interact with a user's session during or before log-in, but not post-login. Therefore improving static risk-based authentication is a practical challenge.

Our proposed solution to address the above mentioned shortcomings of the individual context-based risk assessment techniques is to synergistically consider machine-learning based methods to detect anomalies in both context (browser type, country of origin of IP etc.) and behavioral features of a given user at login time. By considering a model that takes into account several features of browser, operating system, internet connection, connection times, keystroke and mouse dynamics one gains more confidence on the legitimacy of a given log-in attempt. Our model analyzes several previous log-in attempts in order to evaluate the risk of a new log-in attempt and is based on realistic data from customers of several major banks. In summary, the contributions made by this paper are:

- We propose a novel model that combines historical HTTP and behavioral data to detect anomalies during static authentication based part on real data from the banking domain, and part on a publicly available dataset.
- We evaluate the effectiveness of our model obtaining an AUC of 0.957 in our experimental setup.
- We discuss the practical applicability of our solution to realistic industrial scenarios based on our experience in the banking domain.

The rest of the paper is organized as follows: in Sect. 2 we recap some notions of context analytics and behavioral dynamics. In Sect. 3 we present our approach, and describe the data collected and the experimental design. In Sect. 4 we describe the experiments carried out in order to assess the effectiveness of the proposed approach. We discuss related work in Sect. 5 and conclude in Sect. 6.

2   Background and Attacker Model

User authentication has been traditionally based on passwords or passphrases which are meant to be secret. However, secrets can be stolen or guessed and, without further authentication mechanisms, attackers could impersonate a victim and steal sensitive information. To avoid this, the implementation of risk based authentication has allowed traditional authentication systems to increase confidence on a given user's identity by analyzing not only a pre-shared secret, but other features, such as device characteristics or user interaction which are expected to be unique [14,16]. In the following we review some fundamental concepts related to device fingerprinting for authentication and behavioral biometrics.

2.1   Device Fingerprinting

Device fingerprinting is an identification technique used both for user tracking and authentication purposes. The main goal of this technique is to gather characteristics that uniquely identify a device. There are different ways to create this profile, the most reliable of them involves creating an identifier based on hardware signatures. However, acquiring these signatures requires high level privileges on the device, which is often hard to achieve.

Thanks to the popularization of the internet and the increased browser capabilities it is possible to also use statistical identification techniques using information gathered from the web browser [1,18], such as browser history, installed plugins, supported mime types, user agents and also network information like headers, timestamps, origin IP and geolocation. Geolocation can be either collected using HTML5 or approximated from an IP address by using appropriate services. Gathering only browser information means these techniques identify web browsers and not necessarily devices or users. On the other hand, parameters such as HTTP request parameters are easy to spoof. Most recent techniques try to combine both hardware and statistical analysis gathering the information using the web browser capabilities, these techniques use HTML5 [3] and javascript APIs to measure the execution time of common javascript functions and the final result of rendering images as hardware signatures, these measurements are compared to a base line of time execution and rendering performed in a known hardware used as control [5,15].

2.2   User Behavior Identification

Another popular risk-based authentication technique is behavioral analysis, based on mouse and keyboard dynamic statistics. The underlying idea of measuring user behavior is to turn human-computer interactions into numerical, categorical and temporal information. The standard interactions gathered for a behavioral model are key-strokes, mouse movements and mouse clicks. For instance, common features extracted from keyboard events are key pressed and

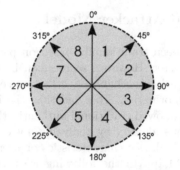

Fig. 1. Mouse directions segmentation.

key released events, together with their time-stamps. For mouse, cursor position, click coordinates and timestamps are commonly used [4]. Such features are processed and aggregated to profile user behavior. In this work, we will use aggregations such as the ones discussed in [7] and summarized in Tables 7 and 8. As shown in Fig. 1 we used the suggested space segmentation in [7] to calculate mouse movement features.

These behavioral features give us information about very unique characteristics of each user such as how fast the user types, how many special keys the user uses, what is the proportion of use of mouse and keyboard, how long the user stops interacting before finishing an activity. The intuition behind this is that it must be easy to distinguish a user who uses mainly mouse from a user who uses mainly keyboard, also intuitively some physical conditions like hardware and user's ability with the peripheral devices makes these interactions more unique.

Behavioral models use machine learning to identify users by using these feature vectors. Notice that by recording one user's interaction in the same situation many times, it is expected that this user will interact with the computer similarly each time and also that it differs from the interactions gathered from other users.

2.3   TWOS Dataset

For the behavioral dynamics analysis, that we will illustrate in the following sections, it is important to have mouse and keyboard dynamics data, in order to evaluate our models. For this purpose, we have chosen to user data from a public data set known as The *Wolf Of SUTD (TWOS)* [4]. The data set contains realistic instances of insider threats based on a gamified competition. We have chosen this dataset since it contains both mouse and keyboard traces, among others. In [4], authors attempted to simulated user interactions in competing companies, inducing two types of behaviors (normal and malicious). The data set contains both mouse and keyboard data of 24 different users. We chose the TWOS dataset because of the large amount of behavioral patterns they recorded. In total, TWOS data set has more than 320 h of mouse and keyboard dynamics.

Data was continuously collected for volunteers during routine internet browsing activities in the context of a gamified experiment. The mouse agent collected the position of the cursor in the screen, the action's timestamp, screen resolution, the mouse action, and user ID. The mouse actions involved in our analysis are mouse movement, button press/release and scroll. The keyboard agent logged all characters pressed by the users. The data set includes the timestamp of event, movement type (press/release), key and user ID. Both alphanumeric and special keys were recorded by the agent. Since the users typed potentially sensitive information the data is provided in an anonymized fashion. The keyboard was divided into zones to accomplish the anonymization. Figure 2 shows the mapping of the keyboard into three zones to enhance the privacy concerns.

$$\{I,O,P,J,K,L,N,M\} \rightarrow \text{RIGHT}$$
$$\{R,T,Y,U,F,G,H,V,B\} \rightarrow \text{CENTER}$$
$$\{Q,W,E,A,S,D,Z,X,C\} \rightarrow \text{LEFT}$$
$$\{0,1,2,3,4,5,6,7,8,9\} \rightarrow \text{DIGIT}$$

Fig. 2. Keyboard mapping layout to anonymize sensitive information.

2.4 Attacker Model

We assume an attacker that has gained access to a victim's credentials to authenticate to a webservice (login and password). An attacker may also gain knowledge about, or try to guess, the context in which a victim uses a service: the time of the day in which a user usually connects, the operating system used, the browser used and IP range from which a victim connects. We assume that an attacker could employ one of the following strategies, or more than one in combination with others to attempt to impersonate a victim:

- *Simple attack:* The attacker connects to the webservice from a machine different than the victim's machine.
- *Context simulation attack:* The attacker connects to the webservice from a machine different than the victim's machine, but tries to replicate or guess the victim's access patterns: OS, Browser type, IP range and time of the day similar to victim's access patterns.
- *Physical access to victim's machine:* An attacker connects from the victim's machine, thereby having very faithfully replicated a victim's context, and attempts at impersonating the victim.

Note that we explicitly exclude from the attacker's capabilities that of recording and attempting to replicate a victim's behavioral dynamics (keyboard and mouse usage features). We believe that although this is an interesting attacker model, it is an extremely powerful one, and we leave its treatment to future work.

3 Approach

The goal of our approach is to overcome the shortcomings of the single risk assessment strategies (context-based analysis of HTTP connections and behavioral dynamics) by obtaining a single model that takes into account both strategies.

In Table 1 we summarize the effectiveness of various strategies in detecting the attacks discussed in the previous section, and also highlight the desired outcome of our approach. In essence, we expect a combined model to perform better in case of attacks, given that the combined model can recognize both changes in context and changes in behavior. Note that in this table we assume there is always impersonation (and thus always changes in behavior).

Table 1. Strategies vs. attack vectors

Approach	Simple attack	Context simulation	Physical attack
Context analytics	Effective	Partially effective	Ineffective
behavioral dynamics	Partially effective	Partially effective	Partially effective
Combination	Effective	More effective than single approaches	Partially Effective

Moreover, we highlight the potential misclassification of the various approaches in various scenarios in Table 2. Here, we summarize the expectation of the combination of both approaches in terms of reducing false positives. When a user uses a new device, one would expect its behavior to be similar in terms of keystrokes and mouse dynamics (although not exact). When he travels, it should remain very similar, those correcting possible false positives from the context analysis.

Table 2. Strategies vs. benign context changes.

Approach	New machine	User travels
Context analytics	Likely FP	Likely FP
Behavioral dynamics	Likely accurate	Accurate
Combination	Likely accurate	Accurate

In the following we will summarize the models we used for the single risk-based strategies, and describe how these models are used in combination to produce a combined risk-based assessment strategy. It is important to note that for the context analytics data we will assume that some users have a heterogenous access pattern (i.e. from multiple devices and locations, due to travel), as depicted in Fig. 3 for a user for which we have 338 access records. On the other hand, the time of activity considered for behavioral interaction reflects the average time of a password based log-in (which typically is a value between 25 and 30 s). Because of these challenges single models are not perfect within a global context attack, but can be used in synergy to produce a better model as we will show in the evaluation section.

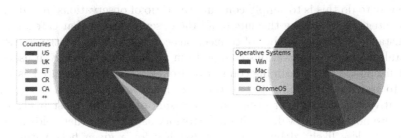

(a) Connection per country distribution.   (b) Connection per operative system distribution.

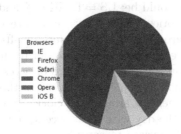

(c) Connection per browser distribution.

Fig. 3. Context of user with heterogeneous access patterns.

3.1   History-Aware Context Analytics

In this subsection we describe the high-level construction of a session context model, based solely on session data obtained from HTTP requests. We assume users with complex access behaviors such as the ones depicted in Fig. 3, so we need to build a system that is good at detecting anomalies and potential attacks, but also it is somewhat flexible to certain changes in context that could be benign.

We assume a system that records usage statistics of the number of times that a user logs in, the day and time of the week at which the user logs in, what type of device and browser they are using, and the country and region from which the user is accessing. Currently, platform and browser data is obtained parsing the user agent, and geographic data is obtained parsing the IP address, information that can be obtained from network sessions corresponding to successful log-ins for a given user.

One of the challenges of building such a model is the fact that several categories considered are non-numerical (for instance a given browser version or operating system). This forces us to use a feature vector with connections statistics on each browser model version, each day of the week, each country and region etc. On the other hand, we must somehow assess the likelihood of a given connection context in order to decide whether a new connection is anomalous or not. One way to do this is to simply compute the ratio of observations in a given field of a category divided by the sum of all the observations in that category.

For instance, let c the number of connections coming from a country K. Let C the total number of observations coming from all countries for a given users. Then the likelihood of an incoming connection from K could be computed as $\frac{c}{C}$. In order to assign a probability of 1 to the most likely event within a category, and a relative weight to other events in decreasing order from most likely to less likely, we normalize all values within a category as follows: order fields from most likely to less likely, define a new probability for a given field within a category as the sum of the probabilities for categories with probability equal or less to the one of the given field. For example, consider three countries with the following probabilities based on access frequency: US $= \frac{1}{2}$, UK $= \frac{1}{3}$, FR $= \frac{1}{6}$. The normalized probabilities would be: US $= 1$, UK $= \frac{3}{6}$ and FR $= \frac{1}{6}$.

Moreover, temporal categories (hours, days, etc.) are considered cyclical, because for instance events around midnight (before or after 24:00) should be considered relatively close to each other. Also, in order to smooth the notion of 'closeness' in discretized events such as frequencies of access in different hours of the day, we use a convolution as depicted in Fig. 4. In this example, we have a distribution of discrete frequencies around the clock for a given user. In this scenario, 7PM is the hour of the day with most access. However this is close to say, 8PM, so it would be appropriate to consider an access at 8PM relatively normal for this context.

The feature vector for a session login attempt is formed using the normalized probability for each variable gathered from the HTTP request. For example, in the countries case above, a session which comes from US will have a value of 1 for variable country in the feature vector. To train the model we calculate the probability profiles for each user using the login history. Afterwards we evaluate a subset of new logins with the user probability profile and compute the feature vector for each visit. The feature vector is fed to a Random Forest model that assesses how anomalous the current event is. The impersonation records were synthesized comparing login events from one user to the history of another user. With this in mind, the model assesses the likelihood of an impersonation. Finally the statistics are updated, the idea being that the system will gradually adapt to permanent changes in the user profile.

3.2 Behavioral Dynamics Combining Keystrokes and Mouse Activity

Both keyboard and mouse events are enough to describe a human-computer interactions and turn it into behavioral features. It is obvious that a regular user uses both at the same time. However, there is no simple way to merge both

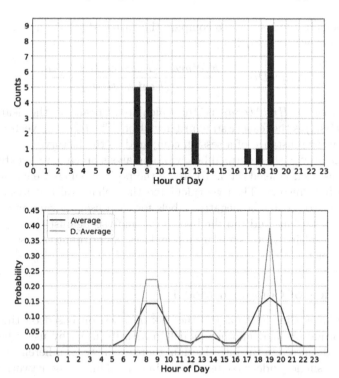

Fig. 4. Graphical representation of convolution used for temporal categories (e.g. hour of connection).

keyboard and mouse dynamics features. To describe a user behavior during a session we calculate the keyboard and mouse dynamics as behavioral features, as described in Tables 7 and 8, using all the gathered events in one single session, where a session is defined as a time frame where the user is performing any activity on the computer. Once the keyboard and mouse dynamics are calculated, we combine both set of features, resulting in only one single vector of features per session. The combination of both set of features describes the use of keyboard and mouse dynamics in a single session. This process is repeated each time a new session is gathered. To compare a session behavior vector against the sessions in history we defined a maximum number of sessions to compare, in our experiment for each user we randomly chose between 10 and 30 sessions, this allows to test the algorithm performance with different history length. We calculated the history mean by using Eq. 1, as follows:

$$FeatureHistMean_j = \frac{\sum_J FeatureHist_j}{|J|} \qquad (1)$$

Where $FeatureHistMean_j$ is defined as the mean of one feature, $FeatureHist_j$ is the individual observation of the feature and J is the number of observations in the history. To compare the gathered session against the user sessions history we used Eq. 2.

$$FeatureDist_i = \frac{Feature_i - FeatureHistMean_i}{\sigma(FeatureHist_i)} \tag{2}$$

Where $FeatureHistMean_i$ is the calculated mean of the feature and $\sigma(FeatureHist_i)$ is the feature standard deviation. The resulting vectors of deviations give us the distance of a session compared to the history.

Using the previously described behavioral analysis process, we created a data set of sessions with labeled data. To create the positive labels we calculated for each user a base history. Then we calculated the behavioral features and deviation vectors. To create the negative labels for each user we randomly selected different users sessions and ran the behavioral analysis against the original user history. The resulting vectors feed a random forest algorithm to assess if a session is legitimate or not.

3.3 Overview of Combined Model

Assume we have a model to assess the risk of a session based on the browser context information, and another model to identify users by using behavioral patterns. As discussed in the introduction, there are however inherent limitations to each of the single models: context-based info of an incoming network (HTTP) connection cannot detect advanced impersonation attacks, whereas behavioral info is not accurate enough in short interactions such as log-ins. As a result we propose to enhance the risk-based authentication system's overall performance by combining the predictions of both models.

In principle, there are several ways to build such a combined meta-model, for instance by building a decision flowchart that takes the scores produced by the singles models and decides whether a given session should be considered suspicious or not. For simplicity's sake, in the following we propose to consider a parametric linear combination of the scores. In the evaluation section we will discuss an example instance of the parameters.

Let us to define $\hat{y}_c, \hat{y}_b \in [0, 1]$ as the prediction of context-based and behavioral model, respectively. We unify the models' prediction using a linear convex combination as we describe in Eq. 3.

$$\hat{y}_t = \alpha_c \cdot \hat{y}_c + \alpha_b \cdot \hat{y}_b \tag{3}$$

where $\alpha_c, \alpha_b \in [0, 1]$ are the coefficient parameters of each model. Note the coefficients must satisfy $\alpha_c + \alpha_b = 1$, because to be a meaningful prediction $\hat{y}_t \in [0, 1]$. As a result we expect, by considering a model that takes into account several learned features of browser context and behavioral dynamics, one gains more confidence on the legitimacy of a given log-in attempt.

Scalability of the combined model Note that the models obtained for the two risk assessment strategies involve training with a dataset of multiple users, however one model is generated that can be applied for each user (there is no need to build one model for each user). Therefore, the approach is designed to scale to millions of users, once the two respective models are trained.

4   Evaluation

In order to train and evaluate the performance of our proposed method we collect two sets of data. The behavioral data set, containing both mouse and keyboard data, was retrieved from a public data set known as *The Wolf Of SUTD (TWOS)* [4] as we describe in Subsect. 2.3. Conversely, the context analytics data set was collected in house from banking web services. This data set contains information about context-based features for online banking log-in sessions. The context-based data set has ca. 13 million entries summarizing connection features when users perform a password-based authentication process. Within those features each entry has information of session timestamp, IP Address and user agent.

For the behavioral dynamics analysis, first we extract mouse traces and keystrokes from the TWOS dataset for all users. The next step is to correlate the mouse and keyboard data for the different sessions the users performed. For each user, a session is created by collecting all mouse entries within a time window before the final login click is observed. Afterwards, the keyboard data is correlated searching for all keystrokes in keyboard data set within the same time windows for the same user. With both mouse and keyboard session's information, we created the features described in Sect. 3. Once the feature data sets are correlated we train the behavioral dynamics model. A random forest was trained in order to capture the behavioral patterns of each user. In order to test the performance of the model we split the full data set into 70% to train and the remaining data's entries to test model. The evaluation of the performance is done using standard classification evaluation measures. Using a confusion matrix, the following measures are calculated:

$$\text{Accuracy} = \frac{TP+TN}{TP+TN+FP+FN} \qquad \text{Recall} = \frac{TP}{TP+FN}$$

$$\text{Precision} = \frac{TP}{TP+FP} \qquad \text{F1-Score} = \frac{2P\Delta R}{P+R}$$

where P, R, TP, FN, TN and FP are the precision score, recall score, the numbers of true positives, false negatives, true negatives and false positives, respectively. We define as positive the sessions with context simulation or impersonation attacks. The sessions without any attack are the negative ones. As we are facing a classification problem some performance metrics are dependent of the decision threshold λ. The λ parameter defines the minimum output probability a prediction must hold to be classified as a attack. In the Table 3 we summarize the performance of the single behavioral dynamics model for different decision threshold.

Table 3. Single behavioral model performance for different classification thresholds (λ)

Decision threshold	F1-score	Precision	Accuracy	Recall
0.3	0.798	0.862	0.725	0.743
0.5	0.735	0.932	0.680	0.607
0.7	0.593	0.972	0.572	0.427

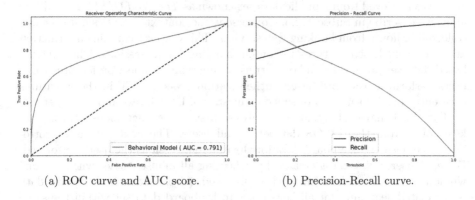

(a) ROC curve and AUC score.　　　　(b) Precision-Recall curve.

Fig. 5. Threshold dependent performance curves for the single model of behavioral dynamics.

From the Table 3 we observe the model has a high precision independent of the decision threshold. However, the cost of high precision is the sensitivity of recall metric while the threshold is increasing. The problem addressed in this paper considers the high cost of false negative predictions because they generates a cascade of attacks which the system does not alert. For this reason we compare precision and recall curves Fig. 5b for the behavioral model to find out the threshold which minimizes the critical cases. Additionally, we show in Fig. 5a the model receiver operating characteristic (ROC) curve performance.

On the other hand, we train a model for the context-based information. Starting from the session timestamp, IP Address and user agent in the session start we calculate the convolutions and probability profiles described in Sect. 3. From the ca. 13 million of session log in attempts, we take the 30% of data to test the models performance and the remaining to train algorithm. The model used to predict the risk of a connection based on contextual information was a random forest. As we did with the behavioral model, we define as positive the sessions with context simulation or impersonation attacks. The sessions without any attack are the negative ones. Table 3 summarizes the performance of the single random forest model trained to alert attacks based on device context information.

The results in Table 4 show the context-based model performs better than the behavioral model. Particularly, we observe that the model has a minor variation

Table 4. Single context-based model performance for different classification thresholds.

Decision threshold	F1-score	Precision	Accuracy	Recall
0.3	0.803	0.948	0.750	0.697
0.5	0.792	0.972	0.743	0.668
0.7	0.771	0.986	0.725	0.633

for the recall metric when the threshold is increased. We compare precision and recall curves Fig. 5b for the context-based model. Furthermore, we show in Fig. 5a the model area under the curve (AUC) metric.

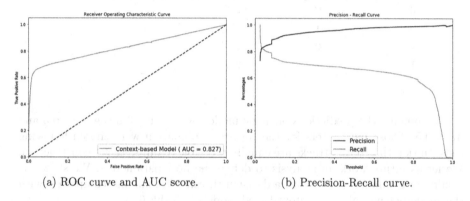

(a) ROC curve and AUC score.　　　　　(b) Precision-Recall curve.

Fig. 6. Threshold dependent performance curves for the single model of context-based analysis of HTTP connections.

The AUC scores for both models are around 0.80, however, the precision and the recall metrics are not accurate enough for the problem we are addressing.

For instance, the recall for context based model indicates a high rate of false negatives which in our context means a high rate of attacks are unnoticeable for the system. Moreover, F1-scores denote that each model separately has a similar performance when they try to detect a global attack. The issue is therefore that each model is not able to detect the counterpart attack: the context-based model will not detect changes in biometric features, and the behavioral model, on its own, will ignore changes in the connection context (Fig. 6).

Therefore we develop a model that attempts to overcome the shortcomings of the single risk assessment strategies (context-based analysis of HTTP connections and behavioral dynamics individually) by proposing a single model that takes into account both strategies, as we describe in Subsect. 3.3. In order to test the combined model we perform a match between the session attempts in TWOS data and context-based data. First we find out the data set with less entries, for us TWOS data. Afterwards we split the TWOS data set into positive (impersonation attacks) and negative samples. As we balanced TWOS dataset before we train the behavioral model the behavioral data has as many positives as negatives entries. We take the positives entries of TWOS and split them into two sets. One of those subsets is matched with an equal number of random sessions from the context-based data set. In that vein, the remaining subset is matched with negatives samples from context-based data. The same process is performed for the positive entries in TWOS data set. As a result, the data set for the combined model is distributed as Table 5 shows.

Table 5. Distribution of combined label to test the model that aggregates the predictions of single models.

Label behavioral	Label context	Data percentage	Combined label
0	0	25%	0
0	1	25%	1
1	0	25%	1
1	1	25%	1

To combine the predictions of single models we use Eq. 3 using $\alpha_c = 0.5$ and $\alpha_b = 0.5$. Those parameters for the convex combination were chosen following the intuition that both attacks are equally probable in our data set construction. We let as future work an analysis to define the best parameters. We show the results for combined model with a decision threshold of 0.3 and compare it with the single behavioral and context-based models in Table 6.

Table 6. Model performance comparison with a decision threshold of 0.3 for the three models we build to increase security in login attempts.

Model	F1-score	Precision	Accuracy	Recall
Behavioral	0.798	0.862	0.725	0.743
Context-based	0.803	0.948	0.750	0.697
Behavioral + Context-based	0.939	0.937	0.910	0.940

The results achieved with the combined model show an important enhancement in detection of attacks, as Table 6 reveals. The high precision and recall values bring to light that the use of a combined model performs better in detecting attacks, given that the combined model can recognize both changes in context and changes in behavior. At the same time, an improved F1-Score and accuracy show that the overall classification was improved, thus also false positives caused by use of new devices or travel can be sometimes mitigated by using the information from the behavioral model (Fig. 7).

Finally, we show in Fig. 8 the model receiver operating characteristic (ROC) for all models we discuss in this work. As it is also evident from the AUC in this figure, a combined model has better performance than the individual models in the data set we have considered.

Scalability of the combined model. We have measure the time it takes to evaluate a given session against the separate strategies, in order to assess the scalability of the approach. These times were measured in a i7-7700hq processor (2.8 ghz), using a single core. For the context-based model, we obtain an execution time of 105 ms in average ($\pm 435\,\mu$s) per session. For the behavioral dynamics model we can classify a session within 106 ms ($\pm 263\,\mu$s) per session. Since the combination of the scores is a simple linear combination, the risk-assessment can be completed within 1 s per each session.

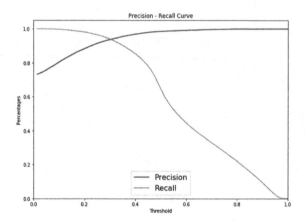

Fig. 7. Precision - Recall curves for the combined risk assessment model (i.e. context-based analysis of HTTP connections and behavioral dynamics)

4.1   Use in Industrial Scenarios

There is no single approach for including an anomaly detection system such as the one discussed in this paper in an operational workflow, nor a one-size-fits-all choice of parameters. Smaller entities – especially if not experiencing a high level of fraud – may want to handle assessments manually. In this case human operators in a SOC receive an alert and react to it. Actions may include blocking the user account, or contacting the user. Aggregated data can also be used to drive the decision process towards more sophisticated, and effective, solutions.

In this scenario where all alerts are handled by a human operator it is mandatory that the alert rate is reasonably small. While it is impossible to define a generic threshold, 1% may be a useful benchmark. As the user base, or the fraud level, grows the institution may decide to integrate the assessments in the application work-flow. If this is done with hard-coded rules then a low rate of false positives is critical, as each alert will generate an action and therefore an expense.

Bigger, or more security-aware, institutions will probably feed the assessment generated from this module to additional systems. While in practice there is no such clear distinction – one single system can often play the two roles – we can typify this systems in two categories:

- *Dynamic authentication systems.* Based on the assessment and other factors such as the user's risk profile and the money at stake the system can decide if additional authentication factors are to be requested to the user, or if access has to be blocked altogether, once or permanently.
- *Transaction anomaly detection systems*, that can include the information related to the transaction to decide if it can be approved, denied or further action should be requested, including sending the transaction to a SOC for further human analysis.

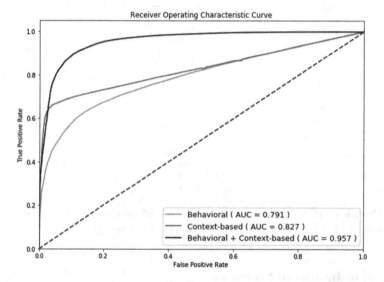

Fig. 8. ROC curves and AUC scores for the single risk assessment strategies (context-based analysis of HTTP connections and behavioral dynamics) and the model which combines both strategies.

In this last scenario a higher rate of false positives is acceptable, as the alerts will be filtered using independent criteria. Furthermore, a numeric assessment is preferable with respect to a binary value, letting the institution fine tune, possibly in real time, how to react to the assessment.

In the context of web-application static authentication, we believe that optimizing the choice of parameters in the model to minimize false negatives (i.e. undetected attacks), is acceptable if in those cases, users can be prompted for a 2-Factor-Authentication such as an OTP sent to their mobile phones. In our model, setting the threshold between 0.2 and 0.25 will yield between 35% and

24.6% false-positive rate against $1,9\%$ and 3.5% false-negative rate respectively. This means roughly one out four users is prompted for 2-FA, whereas between one out 30 to 50 attacks goes undetected. Note that these number hold for our experiments, where 75% of the data consists of attacks, in practice attacks are much less common.

For very sensitive customers, further manual action can be taken depending on the transactions performed in the application. For instance, in the banking domain, further filters depending on transaction amounts can be applied, given a suspicion on context and behavior.

4.2   Discussion and Limitations

We have shown that in principle the combination of both risk-based authentication strategies indeed improves the performance of the single models in isolation. There are a number of limitations to our evaluation. First, the data from the HTTP contextual model and the behavioral model do not belong to the same users. Although in principle there should be no strong correlation between context and behavior, a more accurate model could be built if variations in behavior from the same user across devices are taken into account.

On the other hand, experiments were built under the assumption that the different scenarios combinations (between contextual and behavioral attacks) were equally likely. In practice, attacks are rare, and this aspect should be considered in future work. Last, we have considered behavioral data that has been adapted to simulate static authentication, but that in reality may belong to other activities in the context of the competition where it was gathered. In future work, we plan to consider ad-hoc gathered data from user log-ins. To the best of our knowledge, there is no public database containing both mouse and keyboard data for static authentication, although there are some datasets containing either of them.

Last, regarding Tables 1 and 2, although we have shown implicitly that the overall combined model performs better in both senses (FP and FN), we have not evaluated concretely the single combinations pointed out in the tables, which is also left for future work.

5   Related Work

Risk-based authentication has seen popularity in web applications due to the limitations of password authentication. In [2] Bonneau et al. give a historical overview of the introduction of risk-based authentication in practical systems in order to complement password-based authentication. In [1] Alaca et al. classify and survey several device fingerprinting mechanisms that can be used as the basis for authentication, and discuss different ways in which authentication can be complemented by them. In [8] Misbahuddin et al. study the application of machine learning techniques for risk-based authentication using HTTP and network patterns, in a similar spirit of our technique, but do not take into account behavioral biometric patterns from mouse and keyboard, that as we have shown, improve the accuracy of risk-based authentication.

On the other hand, there are several works exploring applications of behavioral biometrics for static and continuous authentication. In the general context of desktop based applications, Mondal et al. [9] have studied the combination of keyboard and mouse for continuous authentication. Different from them, we focus on static authentication for web applications. In [13] Shen et al. study the applicability of mouse-based analytics for static authentication and conclude that longer than typical log-in interactions would be necessary in order to obtain high accuracy in such models. Traore et al. [16] explore the combination of both mouse and keyboard for risk-based authentication in web applications, however they assume the behavior monitor to be in the application after log-in as well (continuous authentication), and obtain an equal error rate of around 8% (even when considering full web sessions).

To the best of our knowledge the combination of traditional risk-based authentication based on HTTP and network information and behavioral biometrics for static (log-in time) authentication, as proposed in this work, has not been discussed in the literature.

6   Conclusions

The results of our proposed method demonstrates that device identification and behavioral analytics are complementary methods of risk measurement thus by combining both of them, efficacy and performance are never lower than single method approach. Moreover, our approach seams to be more resilient to changes, for instance when a user changes his/her device, an only device identification approach will alert event though there is no attack and an only behavioral approach will not notice the change at all.

In this work we also have shown that, by combining both device identification and behavioral identification risk assessment methods during login time, static web authentication performance can be enhanced by detecting single and mixed attack models with higher or equal accuracy in each case. This also makes web authentication systems more robust and may give the user a better security experience.

We have also discussed the practical applicability of our solution in industrial scenarios. In the future, we plan to consider a more powerful attacker model that is aware of a behavioral risk assessment component and attempts to bypass it, as well as reproducing this experiments on novel datasets that collect both session information and behavioral dynamics simultaneously.

A   Appendix

Table 7. List of behavioral features from keyboard dynamics

Variable name	Variable type	Description
Average Key Press	Float	General time average of key press event
Standard Deviation Key Press	Float	General time standard Deviation of key press event
Median Key Press	Float	General time median of key press event
Count Key Press	Float	General count of pressed keys
Average Key Press per Second	Float	Average time of key press event per second
Average Hold key RIGHT	Float	Average hold time of a right side key
Standard Deviation Hold key RIGHT	Float	Standard deviation of hold time of a right side key
Median Hold key RIGHT	Float	Median hold time of a right side key
Count Hold key RIGHT	Float	Count of pressed keys in the right side
Average Hold key LEFT	Float	Average hold time of a left side key
Standard Deviation Hold key LEFT	Float	Standard deviation of hold time of a left side key
Median Hold key LEFT	Float	Median hold time of a left side key
Count Hold key LEFT	Float	Count of pressed keys in the left side
Average Hold key CENTER	Float	Average hold time of a center side key
Standard Deviation Hold key CENTER	Float	Standard deviation of hold time of a center side key
Median Hold key CENTER	Float	Median hold time of a center side key
Count Hold key CENTER	Float	Count of pressed keys in the center side
Average Hold key ctrl left	Float	Average hold time of a left control key
Standard Deviation Hold ctrl left	Float	Standard deviation of hold time of a left control key

Table 8. List of behavioral features from mouse dynamics

Variable name	Var type	Description
Average Speed Dir1	Float	Average speed in px per second in Direction 1
Average Speed Dir2	Float	Average speed in px per second in Direction 2
Average Speed Dir3	Float	Average speed in px per second in Direction 3
Average Speed Dir4	Float	Average speed in px per second in Direction 4
Average Speed Dir5	Float	Average speed in px per second in Direction 5
Average Speed Dir6	Float	Average speed in px per second in Direction 6
Average Speed Dir7	Float	Average speed in px per second in Direction 7
Average Speed Dir8	Float	Average speed in px per second in Direction 8
Percentage in Dir1	Float	Average speed in px per second in Direction 1
Percentage in Dir2	Float	Average speed in px per second in Direction 2
Percentage in Dir3	Float	Average speed in px per second in Direction 3
Percentage in Dir4	Float	Average speed in px per second in Direction 4
Percentage in Dir5	Float	Average speed in px per second in Direction 5
Percentage in Dir6	Float	Average speed in px per second in Direction 6
Percentage in Dir7	Float	Average speed in px per second in Direction 7
Percentage in Dir8	Float	Average speed in px per second in Direction 8

References

1. Alaca, F., Van Oorschot, P.C.: Device fingerprinting for augmenting web authentication: classification and analysis of methods. In: Proceedings of the 32nd Annual Conference on Computer Security Applications. pp. 289–301. ACM (2016)
2. Bonneau, J., Herley, C., Stajano, F.M., et al.: Passwords and the evolution of imperfect authentication. Commun. ACM **58**, 78–87 (2014)
3. Nakibly, G., Shelef, G., Yudilevich, S.: Hardware fingerprinting using HTML5, pp. 1–13 (2015)
4. Harilal, A., et al.: The Wolf Of SUTD (TWOS): a dataset of malicious insider threat behavior based on a gamified competition. J. Wirel. Mob. Netw. **9**, 54–85 (2018). https://doi.org/10.22667/JOWUA.2018.03.31.054
5. Sanchez-Rola, I., Santos, I., Balzarotti, D.: Clock around the clock: time-based device fingerprinting, pp. 1–13 (2018)
6. Kaspersky: Zeus malware (2019). https://usa.kaspersky.com/resource-center/threats/zeus-virus
7. Bailey, K.O., Okolica, J.S., Peterson, G.L.: User identification and authentication using multi-modal behavioral biometrics. Comput. Secur. **43**, 77–89 (2014)
8. Misbahuddin, M., Bindhumadhava, B.S., Dheeptha, B.: Design of a risk based authentication system using machine learning techniques. In: 2017 IEEE SmartWorld, Ubiquitous Intelligence Computing, Advanced Trusted Computed, Scalable Computing Communications, Cloud Big Data Computing, Internet of People and Smart City Innovation, pp. 1–6 (2017)

9. Mondal, S., Bours, P.: Combining keystroke and mouse dynamics for continuous user authentication and identification. In: 2016 IEEE International Conference on Identity, Security and Behavior Analysis (ISBA), pp. 1–8. IEEE (2016)

10. Newman, L.: Hacker lexicon: what is credential stuffing? Wired Magazine (2019). https://www.wired.com/story/what-is-credential-stuffing/

11. Perrig, A.: Shortcomings of password-based authentication. In: 9th USENIX Security Symposium, vol. 130. ACM (2000)

12. Salem, M.B., Hershkop, S., Stolfo, S.J.: A survey of insider attack detection research. In: Stolfo, S.J., Bellovin, S.M., Keromytis, A.D., Hershkop, S., Smith, S.W., Sinclair, S. (eds.) Insider Attack and Cyber Security. ADIS, vol. 39, pp. 69–90. Springer, Boston (2008). https://doi.org/10.1007/978-0-387-77322-3_5

13. Shen, C., Cai, Z., Guan, X., Wang, J.: On the effectiveness and applicability of mouse dynamics biometric for static authentication: a benchmark study. In: 2012 5th IAPR International Conference on Biometrics (ICB) (2012)

14. Swati Gurav, R.G., Mhangore, S.: Combining keystroke and mouse dynamics for user authentication. Int. J. Emerg. Trends Technol. Comput. Sci. (IJETTCS) 6, 055–058 (2017)

15. Kohno, T., Broido, A., Claffy, K.C.: Remote physical device fingerprinting, pp. 1–13 (2004)

16. Traore, I., Woungang, I., Obaidat, M.S., Nakkabi, Y., Lai, I.: Combining mouse and keystroke dynamics biometrics for risk-based authentication in web environments. In: 2012 Fourth International Conference on Digital Home (2012)

17. Yampolskiy, R.V., Govindaraju, V.: Behavioural biometrics: a survey and classification. Int. J. Biom. 1(1), 81–113 (2008)

18. Cao, Y., Li, S., Wijmans, E.: (cross-)browser fingerprinting via os and hardware level features, pp. 1–15 (2017)

19. Zheng, N., Paloski, A., Wang, H.: An efficient user verification system via mouse movements. In: Proceedings of the 18th ACM Conference on Computer and Communications Security, pp. 139–150. ACM (2011)

Using Honeypots in a Decentralized Framework to Defend Against Adversarial Machine-Learning Attacks

Fadi Younis and Ali Miri[✉]

Department of Computer Science, Ryerson University, Toronto, ON, Canada
{fyounis,ali.miri}@ryerson.ca

Abstract. The market demand for online machine-learning services is increasing, and so have the threats against them. Adversarial inputs represent a new threat to *Machine-Learning-as-a-Services* (MLaaSs). Meticulously crafted malicious inputs can be used to mislead and confuse the learning model, even in cases where the adversary only has limited access to input and output labels. As a result, there has been an increased interest in defence techniques to combat these types of attacks. In this paper, we propose a network of *High-Interaction Honeypots* (HIHP) as a decentralized defence framework that prevents an adversary from corrupting the learning model. We accomplish our aim by (1) preventing the attacker from correctly learning the labels and approximating the architecture of the black-box system; (2) luring the attacker away, towards a decoy model, using *Adversarial HoneyTokens*; and finally (3) creating infeasible computational work for the adversary.

Keywords: Adversarial machine learning · Deception-as-a-defence · Exploratory attacks · Evasion attacks · High-interaction honeypots · Honey-tokens

1 Introduction

Recent years has seen an exponential growth in the utilization of machine-learning tools in critical applications, services and domains. This has led to many service providers now offering their machine-learning products in the form of online cloud services, known as *Machine-Learning-as-a-Service* (MLaaS) [20]. These services allow user to simply use these tools, without the efforts needed to train, test, and fine-tune the underlying machine-learning models. While some more experienced users may still prefer to control how their models are constructed and deployed, to protect their models MLaaS providers typically hide the complex internal mechanisms from most of their users, and simply package the services in non-transparent and obfuscated ways. That is they provide their services in the form of a *black-box* [12,18]. This opaque system container accepts some input and produces an output, but in this system the internal details of the

© Springer Nature Switzerland AG 2019
J. Zhou et al. (Eds.): ACNS 2019 Workshops, LNCS 11605, pp. 24–48, 2019.
https://doi.org/10.1007/978-3-030-29729-9_2

prediction model are hidden from the user. Although hiding the internal mechanisms of models used can provide some protection against insider and outsider attacks, these types of deployments remain susceptible to attacks. For example, an attacker can try to mislead and confuse the MLaaS prediction model, using specifically crafted examples, known as *adversarial examples* [12], leading to a violation of the model integrity [18]. Most proposed defences against these types of attacks aim to strengthen the underlying model by training it against possible expected adversarial malicious inputs. These approaches - such as *Regularization* and *Adversarial Training* [26] - may have limited success, as they do not generalize against newer and more complex adversarial inputs. In this paper, we propose a new defence framework, that can provide an additional layer of protection for MLaaS services. As an example, we show how our framework can be used to defend against malicious attacks on *Artificial Neural Network* (ANN) classifiers. It has been shown that adversarial attacks on these type of classifiers can go undetected [15]. Maliciously crafted adversarial examples can be used to exploit *blind spots* in the classifier boundary space. Exploiting these blind sports can be used to mislead and confuse the learning mechanism in the classifier, post model training, for purposes of violating model integrity.

Our challenge here lies in constructing an adversarial defence technique capable of dealing with different, and possibly adaptive types of attacks. Part of our defence framework utilizes *adversarial HoneyTokens*, fictional digital *breadcrumbs* designed to lure the attacker. They are made conspicuously detectable, to be discovered by the adversary. It is possible to generate a unique token for each item (or sequence) to deceive the attacker and track his abuse. However each token must be strategically designed, generated and deliberately embedded into the system, to misinform and fool the adversary. A major component of our defence framework is focused on designing a decentralized network of *High-Interaction Honeypots* (HIHP), as an open target for adversaries, acting as a type of *perimeter* defence. This decentralized network of honeypot nodes act as self-contained *sandboxes*, to contain the decoy neural network, collect valuable data, and potentially gain insight into adversarial attacks. We believe this can also confound and deter adversaries from attacking the target model to begin with. Other adversarial defences can also benefit by utilizing this framework as an additive layer of security to their techniques to protect production servers where learning models reside. Unlike other defence models proposed in literature, we have designed our defence framework to deceive the adversary in three consecutive steps, occurring in strategic order. The information collected from the attacker's interaction with the decoy model could then potentially be used to learn from the attacker, re-train and fortify the deep learning model in future training iterations, but for now this falls out outside of our scope.

The contributions of this paper are the following:

- We propose an adversarial defence approach that will act as a secondary-level of protecting to *cloak* and reinforce existing adversarial defence mechanisms. This approach aims to: (1) prevent an attacker from correctly learning the classifier labels and approximating the correct architecture of the black-box

system; (2) lure attackers away from the target model towards a decoy model, and re-channel adversarial transferability; (3) create unfeasible computational work for the adversary, with no functional use or benefit, other than to waste his resources and distract him while learning his techniques.
- We provide an architecture and extended implementation of the *Adversarial HoneyTokens*, their designs, features, usage, deployment benefits, and evaluations.

This paper focuses on using honeypots in defending against adversarial attacks against machine learning techniques, and in particular deep learning. For completeness, we have included some background and relevant concepts such as *adversarial examples*, *adversarial transferability* and *black-box learning systems* in the appendix. The rest of this paper is organized as follows. In Sect. 1, we present the role of honeypots in our approach, threat models, attack environments and settings. In Sect. 2, we present our 3-tier defence approach. In Sect. 3 we will discuss the related work, followed by conclusions and future work in Sect. 4.

1.1 Problem Definition

The main goal of this paper to build a decentralized adversarial defence framework against adversarial examples. This level of defence will shield the black-box learning system, using honeypots as one of the primary components of deception in building the framework. This decentralized framework must consist of H high-interaction honeypots. Each of these honeypots is embedded with a decoy target model T_{decoy}, designed to lure and prevent an adversary with adversarial input x from succeeding in causing a mislabeling attack $f(x) = y_{true}$ on the target model T_{target}. Essentially, the framework must perform the following tasks below.

- *task 1* - prevent the adversary from mimicking the neural network behaviour in the learning function *f()* and replicating the decision space of the model. This will be done by blocking adversarial transferability, prevent the building of the correct substitute training model $F(S_p)$ from occurring and the transfer of samples from the substitute model F to the target model T_{target}. This makes it difficult to find a perturbation that satisfies $O\{x + \delta x\} = O\{x\}$, since the target model duplicated is fake.
- *task 2* - the framework must lure the adversary away from the target model T, using deception techniques. These methods consist of using: (1) deployment of uniquely generated digital breadcrumbs (HoneyTokens) TK_n, (2) making the network of honeypot nodes easily accessible (3) set up decoy target models T_{decoy}, deployed inside the honeypots for the attacker to interact with, instead of the actual target model T_{target}.
- *task 3* - create an in-feasible amount of computational work for the attacker, with no useful outcome or benefit. This can be accomplished by presenting the attacker with a *non-convex*, *non-linear* and hard optimization problem,

which is generating adversarial samples to transfer to the remote target model T_{target}, which in this case is a decoy; a decoy of the same optimization problem we saw in the earlier sections:

$$x^* = x + argmin\{z : \hat{O}(x + z) \neq \hat{O}(x)\} = x + \delta_x$$

This strenuous task is complicated further for the attacker because in order to generate the synthetic samples, the attacker must approximate the unknown target model architecture and structure F to train the substitute model $F(S_p)$, which is challenging. Evasion is further complicated as the number of deployed honeypots in the framework increases. Therefore, building this system consists of solving three problems in one, preventing of adversarial transferability[1], deceiving the attacker and creating immense computational work for adversary targeting the system to waste computational time and resources; all the later, while keeping the actual target model T_{target} out-of-reach.

The adversarial examples generated need to have such an effect on the classifier, that it explicitly lowers the confidence on the target label. Misclassification attacks, to us, were less attractive since they do not make for interesting adversaries, not to mention the fact that these type of attacks appear random in nature, focusing on an arbitrary set of data samples. With no fringe inconsistencies to dispute, it becomes difficult to discern failures brought about by non-malicious factors effecting the classifier. Building on the latter, misclassification attacks make it all the more difficult to design defences and robust frameworks to thwart adversaries when the attack itself seems arbitrary in nature.

1.2 Threat Model

Attack Specificity - generally, for an adversary to succeed in his attack, and whether the attacker has his sight set on violating the *availability* or *integrity* of the model, adversarial transferability needs to be successful. For purposes of our paper, we have decided to design our adversarial attack to be a *targeted exploratory* one in nature [9]. A targeted attack is when the adversary has a specific set of data samples in mind, and is discriminatory in his attack. This means the adversary wants to force the DNN to output a specific target label y_{target}, $f(x) \longrightarrow y_{target}$, instead of the correct label y_{true}, $f(x) \nrightarrow y_{true}$. See Fig. 1 for an illustration of an adversarial targeted attack, violating model integrity.

Exploited Vulnerability - the cogent properties of adversarial examples x^* make them a prime candidate for adversarial attacks on deep learning systems. It should be anticipated that an ambitious and equally resourceful adversary will conspire to use these perturbations for malicious purposes. Generally, deep

---

[1] Even with the little knowledge possessed by a potential adversary, a targeted attack in a black-box setting is still in fact probable.

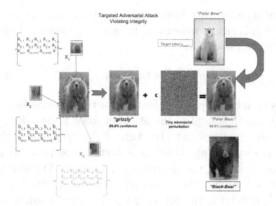

Fig. 1. Input x (left), modification $\delta + x$ controlled by ε (middle) which controls the magnitude of modification in the image, generating the adversarial evading sample x^*(right). As you can see, both bus images look astoundingly similar.

neural nets (DNN) work by extracting and learning the key multi-dimensional discriminate features $X_{m,n} = \{x_{n,1}, x_{n,2}, x_{n,3}, ..., x_{n,m}\}$ embedded within the input sample x pixels, to correctly classify it with the correct output label y_{true}. However, with adversarial examples entities, the acuity of a DNN's classification ability becomes slightly manipulable, and the adversary is aware of this weaknesses.

In our paper, the designed adversary's attack depends on the successful exploitation of a fundamental vulnerability found in most, if not universally all DNN learning systems. This vulnerability is acquired during faulty model training. This weakness is embodied by a lack of non-linearity in poorly trained DNN models, that these visually indistinguishable adversarial examples, born in a high-dimensional space, epitomize. Other factors may also be responsible, such as poor model regularization. This inability to cope with non-linearity makes the DNN classifier insensitive to certain blind-spots in the high-dimensional classification region. Knowing the latter, an adversary can generate impressions of the input samples with slight perturbations. These examples can then be transferred between adjacent models, due to the cross-model-generalization property which allow the transfer of adversarial examples between the original and target model the adversary desires to exploit. The above vulnerability is manifested after the examples are synthesized and injected during the testing phase.

Attacker Capabilities - each honeypot node in the decentralized defence framework contains a decoy target model T_{decoy}, presented to the adversary as the legitimate target model. Here, an *Oracle O* represents the means for the adversary to observe the current state of the DNN classifier learning by observing how a target model T_{target} handles the testing sample set $(x^{'}, y^{'})$. In our attack environment, querying the *Oracle O* with queries $q = \{q_1, q_2, q_3, ..., q_n\}$ is the exclusive and only capability an adversary possesses for learning about the target model and collecting his synthetic dataset S_p to build and gradually train his DNN substitute model F.

The adversary can create a small synthetic set of adversarial training samples from the initial set S_0 with output label y' for any input x' by sending $q_n > 1$ queries to the *Oracle O*. The output label y' recurred is the result of assigning the highest probability assigned a label y' which maps back to a given x' is the only capability that the attacker has for learning about presumed target model T_{target} through its *Oracle O*. The attacker has virtually no information about the DNN internal details. The adversary is restrained by the same restrictions a regular user querying the *Oracle O* has. The latter is something an adversary should adhere to make his querying attempts seem harmless, while engaging the decoy model within the adversarial honeypot. Finally, we anticipate that the adversary will not restrict himself to querying one model and will likely connect to multiple nodes and DNN model classifiers from the same connection for purposes of synthetic data collection in parallel. This should trigger an alarm within our framework, indicating multiple access and that something abnormal is occurring.

1.3   Attack Setting

Our envisioned profile for the adversary targeting our black-box learning system does not possess any internal knowledge regarding the core functional components of the target model T_{target} DNN. This restriction entails no access to model's *DNN architecture, model hyper-parameters, learning rate, etc.* We have already established that an adversary can prepare for an attack by simply monitoring target model T_{target} through its *Oracle O* and use the labels to replicate and train an approximated architecture F.

The ad-hoc approach at the adversary's disposal is that he can learn the corresponding labels by observing how the target model T_{target} classifies them during the testing phase. The adversary can then build his own substitute training model F and use this substitute model F in conjunction with synthetic labels S_p to generate adversarial examples propped against the substitute classifier, which the attacker has access to. Even if the substitute model S and target model T_{target} are different in architecture, the adversarial examples x^* generated for one can still tarnish the other if transferred using *adversarial transferability*. Since the adversarial examples between both models are only separated by added tiny noise ε, the examples look similar in appearance. The latter is true even if both models, original T_{target} and substitute model F, differ in architecture and training data. As long as both models have the same purpose and model type. Although the *Adversarial transferability* phenomena is discouraging, but alone it is advantageous for the adversarial attackers to launch targeted attacks, with little or no constraint on their attack blueprint. Adversarial transferability eventually becomes a serious concern because attacks will grow in sophistication and potency over time. It is challenging to design a model that can generalize against more advanced attacks, if not all. Also, it is difficult to dismantle and reverse-engineer how these attacks propagate and cause harm, since no tools exist to expedite the process to learn from the attack in time to re-train the network.

2   Deception-As-A-Defense Approach

The proposed *Adversarial Honeynet* framework is considered as an added layer of protection to *blanket* a deployed deep learning system, in order to combat imperceptible adversarial examples, within a black-box attack setting. There are several advantages and benefits that this framework can bring in the protection of existing learning systems. A single adversarial honeypot node in this decentralized framework may offer the following benefits: (1) *adversarial re-learning*; conceptually, it is a pragmatic method of collecting intelligence on the adversary, such as attack patterns, propagation, frequency and evolution. The latter results can be used to learn and reverse-engineer adversarial attacks; (2) an *anomalous classifier* used to identify whether the attackers actions are malicious or benign, this will help to determine whether or not to record the attackers session information based on behaviour patters against a *white-list*; (3) a *decoy target model*, used as a placeholder for the adversary to engage and interact in case his intention are indeed malicious in nature. The attacker's interaction with model is represented by the *Oracle* $\hat{O}$, that an adversary observes and queries, re-channeling his efforts; (4) an *Adversarial Database*, used to collect and securely store attack session data on the adversary's actions and maneuvers, used later to research and understand the adversary in *adversarial re-learning*.

2.1   Adversarial Honeynet

All honeypot nodes are deployed with identical decoy models T_{decoy} that resemble the original target DNN model T_{target}. Also, all services and applications on the high-interaction honeypot are real and not simulated, prompting the attacker to assume the model is indeed real, published or leaked by mistake. Neighbouring adversarial honeypots are called *HoneyPeers*, these nodes are always active and have a weak non-privileged TCP/IP port open that is known to attract adversaries, supported with adversarial honey-tokens. The docker container node begins recording information when the anomalous classifier detects that the attacker is attempting to do something malicious and discretely notifies the neighbouring *HoneyPeers* that an attacker is active within the network. *HoneyCollectors* are used to aggregate and collect information from each individual adversarial honeypot node and store it in the central *Adversarial Database*. All activities on the node are collected and stored with a public-key hashed *timestamp*. In our framework, the central database is a *Samba* database is used to collect *structured, unstructured*, and *semi-structured* session data to record the *adversary-honeypot-decoy* interaction. An analysis module, used to aggregate adversarial information and use that to learn about the attacker, this learned information can potentially be used to perform inference for future attacks. Figure 2 gives an illustration of our adversarial honeynet architecture.

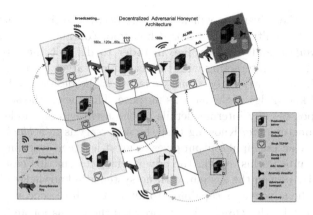

Fig. 2. Adversarial honeynet architecture

2.2   Honeynet Functional Components

- **HoneyPeers** are a series of interconnected high-interaction honeypots joined
  in a decentralized network topology. Each HoneyPeer is an autonomous high-
  interaction honeypot contained node, with a copy of the decoy learning model
  T_{decoy}, embedded within a monitored Linux container, powered by Docker.
  Encrypted communication messages are passed between the nodes in order
  to notify adjacent nodes that an attack is occurring or has occurred. All communication is governed by our *message-passing-protocol* defined in the next
  section. Each *node-to-node* interaction is initiated by exchanging a *HoneySession Key*, which is used to authenticate a node's identity with each of its peers
  and is reused in verify future interactions. If a node should become unrespon-
  sive, it is assumed that the node has been *compromised and is infected*. In the
  case that a node should become infected, it can be assumed has been compromised by the adversary, in which case all neighbouring nodes will sever all
  future communication with it, flag any local session HoneySession keys, and
  the infected honeypot will be cautionary labeled. Furthermore, all node-to-
  node interactions are securely stored and recorded in the *central adversarial
  database*.
- **Decoy Classifier** represents our solution for preventing the adversary from
  interacting with the target classifier learning model T_{target}, and block *transferability* from occurring by re-channeling it to the honeypot. We distribute
  fake *decoy* learning systems throughout the enterprise or specifically in the
  anterior of a production system, acting as a type of *sentinel*. In this paper,
  we hypothesize that legitimate users querying the learning system have no
  cause to interact with decoys or take notice of our adversarial honeypot.
  We decided to experiment with *deception-as-a-defence* using honeypot and
  decoys because we wanted to give the adversary a false sense of assurance,
  then identify and study them, and greatly reduce the rate of false-negatives
  FN violating classifier integrity.

We suspect the adversary will attack our decoy learning classifier system T_{decoy} once he infiltrates the tailored honeypot container. It's purpose is to simply *simulate* and *mimic* value, in order to distract the adversary and prevent him from interacting with the legitimate target model T_{target}. If we consider the adversary to be *weak*, we see that the designed adversary only has partial knowledge of the model's purpose. This means the adversary does not have possess any internal details of the architecture, hidden layers, or hyper-parameters, etc. Knowing that the adversary is in a *black-box* setting and can only access input/output gives us great leverage over him. Before the adversary launches his attack, the adversarial actor in this case is like any other regular user in the system, with no systematic knowledge of the classifier. Here, the adversary's capability to interact with the decoy model T_{decoy} is represented by the *Oracle* $\hat{O}$. $\hat{O}$ represents the means for an adversary to interact with and learn from decoy model. Since the adversary wishes to produce adversarial examples x^* for a specific set of input samples $\bar{x}$, collected by querying the $\hat{O}$, and then transfer them. However, adversarial transferability can be *re-channeled* if we can switch the target model T_{target} and the *Oracle* O with a decoy model T_{decoy} and thereupon *Oracle* $\hat{O}$, and convince the adversary that no *tampering* has occurred.

- **HoneyCollector** is the component responsible for collecting all the adversarial session information on the adversary within each of the honeypot nodes in the network, it is the *Samba* component within our system.
- **Anomaly Classifier** used to predict whether the adversary's actions inside the honeypot are considered abnormal or not. It depends on indicators, such as (1) *Number of DNN labeling requests*; (2) *execution of unusual scripts*; (3) *irregular outbound traffic from source*; (4) *sporadic DNN querying*; (5) *persistent activity on the DNN*; (6) *use of foreign synthetic data for labeling*.
- **Adversarial Tokens** - They can be thought of as a *digital* pieces of information. It can manifested from a document, database entry, E-mail, or a credentials. In essence, it could be anything considered valuable enough to lure and bait the adversary.

2.3   HoneyPeer Node Inter-communication

This section describes the message passing protocol between the nodes in the adversarial Honeynet framework. A message can only be sent and received between two HoneyPeer nodes in the network that have exchanged *HoneySession* key between them. Any message that has been received or sent spontaneously should not be accepted. A reliable message passing technology must be set in place to avoid congestion and bottleneck at one of many parts of the network. Also, all messages sent, received, and dropped are *time-stamped* and recorded within the *adversarial central database* for bookkeeping purposes.

- **HoneyPeerALRM** - a distress message indicating that host node (*Sender*) has been compromised. The message is broadcast to the nearest adversarial honeypot node in the network. The neighbouring nodes (*Receivers*) are responsible for intercepting and passing the message to all neighbouring nodes in the network. For obvious security concerns and as fault-resistance, another HoneyPeerALRM message is sent on behalf of the *anomalous classifier*, in the case an adversary manages to seize control of the node and hijack it after detection. Each HoneyPeerALRM message must receive an *HoneyPeerACK* to indicate that the distress HoneyPeerALRM message has been received and acknowledged. Failure to reply might indicate one or several neighbouring nodes have also been compromised. To add, nodes should not receive unsolicited HoneyPeerALRM reply messages from other adversarial nodes, as this may indicate malicious misrepresentation.

- **HoneyPeerAck -** this is a message sent corresponding to each HoneyPeer-ALRM message sent on behalf of the node. A HoneyPeerACK indicates that the distress HoneyPeerALRM message has been received and confirmed by the endpoint node. Failure to receive and acknowledge one ore more *Acks* might indicate that one or all the surrounding neighbouring nodes have been compromised. Also, nodes should not receive unsolicited HoneyPeerALRM reply messages from other adversarial nodes.

- **HoneyPeerSafePulse -** Periodically, a honeypot node will send a *pulse* indicating that it is still active and part of the decentralized network, and not compromised. If the node neighbouring it does not reply in 180 s with an *HoneyPotSafeAck* response, it is assumed that the node has been compromised.

- **HoneyPeerSafeAck -** A confirmation message sent to indicate that the node is active. After 3 consecutive (60 s interval) *no replies*, it can be assumed that either the receiving node is down or has been compromised, in which case, all neighbouring nodes will sever all communication with it, purge any HoneySession keys, and the infected honeypot will be labeled as an *InfectedPeer*.

- **HoneySession Key -** An adversarial session key is exchanged between two HoneyPeer nodes. This HoneySession Key is exchanged at the beginning of a *node-to-node* interaction and will be used an authentication method in future *node-to-node* communications.

2.4   Attracting the Adversary

Adversarial Honey-Tokens. We extended the honeybit token generator in [1] to create the *adversarial honey-tokens* generator, which acts as an automatic monitoring system that generates adversarial deep learning related tokens. It is composed of several components and processes. In order to understand how the system functions, one must have an understanding of the individual operative components and processes. The following points offer an insight into how the system functions used to create token and decoy digital information to bait the adversary:

- **Baiting the Attacker -** in order for the digital tokens generated by the application to bait the attacker successfully they should have the following properties: (1) be simple enough to be generated by the adversarial honey-tokens application, (2) difficult to be identified and flagged as a bait token by the adversary, (3) sufficiently pragmatic to pass itself as a factual object, which makes it difficult for the adversary to discern it from other legitimate digital items. The purpose of these monitored (and falsified) resources is to persuade and lure the adversary away from the target DNN model T_{target}, and bait him to instead direct his attack efforts towards a decoy model T_{decoy} residing within the honeypot trap. The goal here is to allow the adversary's malicious behaviour to compromise the hoaxed model, preventing the adversarial examples transferability to the T_{target} model from occurring, and forcing the attacker to reveal his strategies, in a controlled environment. The biggest challenge associated with designing these tokens is adequate camouflaging to mimic realism, to prevent being detected and uncloaked by the adversary.

– **Adversarial Token Configuration -** the configuration of the adversarial honeypot generator occurs within the .*yaml* markup file (hbconf.yaml). Here, the administrator sets the honeypot decoy *host IP address*, *deployment paths*, and *content format.* The configuration file, through the path variables, set where the tokens will be leaked inside the operating system, offering by that a large degree of freedom. Also, the administrator can customize the individual file tokens, as well as the general honey-tokens and the adversarial machine learning tokens added. As mentioned, this file allows the building of several types of tokens. The first type of tokens are the *honeyfiles*, which include *txtmail*, *trainingdata*, and *testingdata*. These type of tokens are text-based and derive their formatted content from the template files stored in the templates folder. The second type of tokens include network honeybits, which include fake records deployed inside the UNIX configuration file or any arbitrary folder. The latter include general type tokens such *ssh*, *wget*, *ftp*, *aws*, *etc.*, These tokens usually consist of an IP, Password, Port, and other arguments. The third type of tokens deployed are the custom honey-tokens which are deployed in the bash history; these tokens are much more interesting since they take any structure or format the defender desires.

– **Token Leakage -** the most dominant feature of the adversarial honey-token generator is its ability to inconspicuously implant artificial digital data (credentials, files, commands, etc) into the productions server's file system. The embedding location can be set inside the .*yaml* configuration file (hbconf.yaml) using the PATHS: *bashhistory*, *awsconf*, *awscred* and *hosts.* After the defender compiles and builds the adversarial tokens they are stealthily deployed at set path/locations within the designated production server's operating system. There, the tokens reside until they are found and accessed by the adversary. The Docker container at this point records intelligence on the attacker's interaction with the token.

– **Docker to Monitor the Adversary Access -** Docker was selected since it provides a free and practical way to contain application processes and simulate file system isolation, where the adversarial tokens application image will be run. In our defence framework, the numerous production servers not open to the public domain will be reserved for adversarial research to capture intelligence and analyze attacks. They will open via an exposed TCP/IP port open to the public, with weak non-privileged access points. The docker container will act as the *sandbox*, acting as entire layer to envelop the honeytoken application image. Using the insight gained from the adversaries later lured to the honeypots will be used study emergent adversarial strategies, input perturbations and discovering techniques used by adversaries in their exploits. Docker will create a new container object for each new incoming connects and set up a *barrier* represented as the sandbox. An unsuspecting attacker that connects to the container and finds the tokens is presumably lured to the honeypot containing the decoy DNN model T_{decoy}. If the adversary decides to leave, he is already keyed to that particular container using his IP address, which connects him to the same container if he decides to disconnect and then reconnect.

- **Adversarial Token Generation -** through the extended adversarial token framework we compile the tokens using *go build* command. The following are only some of the tokens that can be generated using the adversarial honeytokens framework: (1) *SSH token*, (2) *host configuration token*, (3) *ftp token*, (4) *scp token*, (5) *rsync token*, (6) *SQL token*, (7) *AWS token*, (8) *text-mail token*, (9) *training-data token*, (10) *testing-data token*, (11) *comment tokens*, (12) *SSH password token*, (13) *start-cluster node token*, (14) *prepare DNN model token*, (15) *train DNN model token*, (16) *test DNN model token* and (17) *deploy DNN model token*.

2.5 Detecting Adversarial Behaviour

One of the greatest challenges in this paper was deciding how to adequately detect, classify and label adversarial behaviour as malicious. Not to mention building the actual classification model that would be responsible for doing so would have been a great undertaking on its own. However, there were other practical detection methods at our disposal, such as using signature-based detection to compare an object's behaviour against a *blacklist*, and anomaly-based detection to compare an object against a *white-list*. We chose to lean towards the former method (white-list) over blacklisting since we did not have reliable adversarial data that could have been used to generate a signature to fingerprint a potential adversary. White-list detection works best when attempting to detect entity behaviour that falls out of anticipated and well-defined user actions, such as *over-querying the DNN model*, or *causing a sudden decline in the classification model performance*. White-list based anomaly detection fits perfectly into our defence framework since we can characterize any pattern of activities deviating from the norm as an intrusion. The latter is in our favour since we are trying to detect actions to exploit the classifier which are novel in nature.

2.6 Adversarial Behaviour

In order detect adversarial anomaly behaviour, we have summarized a list of adversarial actions and indicators that may signal an *an-out-of the-ordinary* on the learning model. We will later use this indicators to build our white-list security rules. The following are some of those indicators:

- **Persistent DNN Querying -** While normal (non-adversaries) users will be querying the DNN T_{decoy} model with 1 or 2 queries per session, the adversary will be sending hundreds, if not thousands per session. All this in effort to build his synthetic training dataset S_p, the adversary will need to continuously collect training data, augment it and gradually train his substitute adversarial model $F(S_0)$. Repetitive queries $\tilde{Q}$ from the same source user within a set unit of time might indicate the adversary is query-thrashing the DNN model for labels (x', y'). The latter could be a possible indication of an adversarial attack on the prediction model.

- **Spontaneous DNN Activity -** In order for the adversary to craft adversarial examples x^*, he will need to collect an initial set of labels S_0 from labeling (x', y'). Then, he needs to build a substitute training model F that mimics the learning mechanism inherent in the decoy model T_{decoy}. Naturally, collecting enough sample labels to accurately train the model F requires a large number of queries $\tilde{Q}$ solicited from the *Oracle* $\tilde{O}$. Consequently, in order to avoid raising suspicions, the adversary will try to build this initial substitute model training set S_0, as quickly and discretely as possible. The latter could be a possible indication of an adversarial attack on the prediction model. This is true since a few queries are within normal user behaviour, who have no malicious intent in mind. But spontaneously querying the oracle falls out of normal activity.

- **High number of DNN Labeling Requests -** an abnormally high number of query requests to the Oracle $\tilde{O}$ is not normal either. Let us not forget, that training of the substitute model $F(S_0)$ is repeated several times in order to increase the DNN model accuracy and similarity to T_{decoy}. With each new substitute training epoch e, the adversary returns to $\tilde{O}$ and queries to augment (enlarge) the substitute model training set S_0 produced from labeling. This will produce a large training set with more synthetic data for training. With the correct model architecture F, the enlarged dataset is used to prototype the model's decision boundaries separating the classification regions.

- **Sudden Drop in Classification Accuracy -** Building on the above and as discussed in Sect. 2, our designed adversary seeks to cause a misclassification attack on the target decoy model T_{decoy}, by inserting malicious input in the testing phase. Because of this, an input unrecognizable to the model's discriminate function can be classified with high-confidence *(false positive)*, and an input recognizable to the model can be classified with low-confidence *(false negative)*, violating the integrity of the model. Other factors may influence a drop in accuracy, such as a poor learning or added bias in the data. This does not normally occur in a production environment, which indicates that our classification model is under attack.

  - Other known indicators are more network related, such as *execution of unusual scripts alongside the DNN, Irregular outbound traffic or source, any sensitive or privileged path accessed during the interaction,* and *any spawning of suspicious child process.*

3   Related Work

The literature review below focuses directly on the concept of defending against adversarial examples, aimed at misleading the classifier. Most of the known defence methods are mainly based on data pre-processing and sanitation techniques, employed during the training phase of DNN model preparation. Pre-processing and sanitation typically mean influencing the effect that sample training-set data, X, has on neuron weights of the underlying DNN model, by distinguishing and filtering out malicious perturbations, inserted by an adversary that may mislead and/or confuse the classifier causing a misclassification or violation of model integrity. Other notable work in this section focus on the role of cybersecurity defence through means of deception, specifically with the use of *decoys* and fake entities to deceive the attacker. Our challenge here lays in constructing a secondary-level of protection and defence, designed not to replace known adversarial defence techniques, but to supplement and reinforce existing ones, with the use of adversarial deception re-enforcing the application perimeter.

[27] focuses on addressing the lack of efficient defences against adversarial attacks that undermine and then fool deep neural networks (DNNs). The need to tackle this issue has been amplified by the fact that there is no unified understanding of how or what makes these DNN models so vulnerable to attacks caused by adversarial examples. The authors propose an effective solution which focuses on reinforcing the existent DNN model and making it robust against adversarial attacks, attempting to fool it. The proposed solution focuses on utilizing two strategies to strengthen the model, which can be used separately or together. The first strategy is using a bounded ReLU activation function,$f_R(x) \rightarrow y$, in the DNN architecture to stabilize the overall model prediction ability. The second is based on augmented *Gaussian* data for training. Defences based on data augmentation improve generalization since they consider both the true input and its perturbed version. The latter enables a broader range of searches in the input, then say adversarial training, which is limited in its partial of the input, causing it to fall short. The result of applying both strategies results in a much smoother and more stable model, without significantly degrading the model's performance or accuracy.

Work in [8] is the most relevant academic paper, with regard to motivation and stimulus for the purpose of developing our proposed auxiliary defence technique, using honeypots. The authors in [8] propose a training approach aimed at building adversarial-resistant black-box learning systems against adversarial perturbations, by blocking transferability. The proposed method of training, called *NULL-labeling* works by evaluating input x and lowers confidence on the true label y, if x is suspected to be perturbed and rejecting it as invalid input. The criteria on which the method evaluates x is if it spans out of the training-data data distribution area. The training method smoothly labels, filters out, and discards invalid input (NULL), which does not resemble training-data. This is to prevent from allowing it to be classified into intended target label. The ingenuity of this approach lies in how it is able to decisively distinguish between clean and malicious input. NULL labeling proves its capability in blocking adversarial transferability and resisting the invalid input that attempts to exploit it. The latter is achieved by mapping malicious input to a NULL label and allowing clean test data to be classified into its original true label, all while maintaining prediction accuracy.

In [21], a training approach for combating adversarial examples and fortifying the learning model. The authors propose this defence technique in response to adversarial examples, with their abnormal and ambiguous nature. The authors argue that model adversarial training still makes the model vulnerable and exposed to adversarial examples. For this very purpose, the authors present a data-training approach, known as *Batch Adjusted Network Gradients* or *BANG*. This method works by attempting to balance the causality that each input element has on the node weight updates. This efficient method achieves enhanced stability in the model by forming *smoother* areas concentrated in the classification region that has classified inputs correctly and has become resistant against malicious input perturbations that aim to exploiting and violating model integrity. This method is designed to avoid instability brought about by adversarial examples, which work by *pushing* the misclassified samples across the decision boundary into incorrect classes. This training method achieves good results on DNNs with two distinct datasets, and has low computational cost while maintaining classification accuracy for both sets.

In [2], the authors suggest a framework that actively and purposefully leaks digital entities into the network to deceive adversaries and lure them to a honeypot that is covertly monitors, tracks token access, and records any new adversarial trends. In a period of one year, the monitored system was compromised by multiple adversaries, without being identified as a controlled decoy environment. The authors argue that this method is successful, as long as the attacker does not change his attack strategy. However, a main concern for the authors is designing convincing fake data to deceive, attract, and fool an adversary. The authors also argue that the defender should design fake entities that are attractive enough to bait the attacker, while not revealing important or compromising information to the attacker. The defender's goal is to learn as much as possible about the attacker. The message that the authors try to convey is that as the threat of adversarial attacks increases, so will the need for novelty in the defence approaches to combat it.

Work in [19], serves as an examination of the concept of *fake entities* and digital tokens, which my framework partially relies upon. Fake entities, although primitive, are an attractive asset in any security system. The authors suggest fake entities could be *files, interfaces, memory, database entries, meta-data, etc.* For the authors, these inexpensive, lightweight, and easy-to-deploy *pawns* are as valuable as any of the other security mechanisms in the field, such as firewalls or a packet analyzers. Simply, they are digital objects, embedded with fake divulged information, intended to be found and accessed by the attacker. The authors advocate that operating-system based fake entities are the most attractive and fitting to become decoys, due to the variety of ways the operating system interface can be configured and customized. Once in possession of the attacker, the defender is notified and can begin monitoring the attacker's activity. Later in this work, the authors implement a framework that actively leaks credentials and leads adversaries to a controlled and monitored honeypot. However, the authors have yet to build a functioning proof-of-concept.

There is also extensive work done on utilizing adversarial transferability in other forms of adversarial attacks, deep learning vulnerabilities in DNNs, and black-box attacks in machine learning. Among other interesting work that served as motivation for this paper include: utilizing honeypots in defence techniques, such as design and implementation of a honey-trap [5]; deception in decentralized system environments [22]; and using containers in deceptive honeypots [11]. Our approach using honeypots, does not seek to replace any of the existing methods to combat adversarial examples in a black-box attack context. However, it can be used effectively as an auxiliary method of protection that strengthen existing defence methods in production systems.

4   Conclusions

In this paper, we have discussed adversarial transferability of malicious examples, and proposed a defence framework to counter it, using deception derived from existing cyber-security techniques. Our approach is the first of its kind to use methods derived from cyber-security deception techniques to combat adversarial examples. We have shown it to be possible to use deception to prevent an adversary from mimicking a target model's classification behaviour, if we successfully re-channel adversarial transferability. We have also presented a novel defence framework that essentially lures an adversary away from the target model, and blocks adversarial transferability, using various deception techniques. We proposed presenting the adversary with an infeasible amount of computational with no useful outcome or benefit. This can be accomplished by presenting the attacker with a hard *non-convex* optimization problem, similar to the one used for generating adversarial samples. Our framework allows the adversary to transfer these examples to a remote decoy learning model, deployed inside a high-interaction-honeypot.

A   Appendix

A.1   Deep Neural Nets (DNNs)

Deep Neural Networks (DNNs) are a widely known machine-learning technique that utilizes n parametric functions to model an input sample x, where x could be an image tensor, a stream of text, video, etc. [18]. DNNs differ from conventional neural networks is the large number of (hidden) learning layers they can use, which in return allows these models to adapt to intricate features and solve complex problems. Amongst the countless uses for DNNs' is their utility in building image classification systems that can identify an object from the its intricate edges, features, depth and colours. All of that information is processed

in the hidden layers of the model, known as the *deep* layers. As the number
of these *deep* layers increases, so does the capability of the DNN to model and
solve complex tasks. Simply expressed, a DNN is composed of a series of parametric functions. Each parametric function f_i represents a hidden layer i in
the DNN, where each layer i compromises a sequence of *perceptrons* (artificial
neurons), which a processing units that can be modeled into chain sequence of
computation. Each neuron maps an input x to an output y, $f : x \longrightarrow y$, using
an *activation function* $f(\varphi)$. With each layer, every neuron is influenced by a
parameterized *weight vector* represented by θ_{ij}. The weight vectors holds the
knowledge of the DNN when it comes to training and preparing the model F. A
DNN computes and defines model a F for an input x as follows [18]:

$$F(x) = f_n(\theta_{ij}, f_{n-1}(\theta_{n-1,j}, \cdots, f_2(\theta_{2j}, f_1(\theta_{1j}, x))))$$

A.2 Security of Deep Learning

In recent deep learning literature, there has been a lot of works that has focused
on deploying deep neural networks in malicious environments, in which the network is potentially exposed to numerous attacks [6, 12, 26]. At the centre of these
threats are *Adversarial Examples*. Adversarial examples are *perturbed* or modified versions of input samples x, that are used by adversaries to mislead and
exploit deep neural networks, during test time, after training of the model is
completed [16]. They are injected in order to circumvent the learning mechanism acquired by the DNN with the goal of misclassifying a target label. They
are crafted with carefully articulated perturbations, added to the input $x + \delta x$,
that *forces* the DNN to display a different behaviour than intended, chosen
by the adversary [16]. It is important to note that the magnitude of perturbations must be kept *small enough* to have a significant effect on the DNN, yet
remain unnoticed by a human being. These adversarial exploitations vary in
their motivation for corrupting a DNN classifier, however some of the most common incentives range from simply reducing the confidence of a target label to a
arbitrary source-label misclassification [16]. Confidence reduction entails reducing the accuracy on a label y for a particular input x in the testing pair (x', y').
By contrast, source label misclassification involves having the model classify an
input x as a chosen target label y_{target}, different from the original (and intended)
true source label y_{true}. For any attack to be successful, it requires the adversary to have previous knowledge of the DNN architecture, preferably a strong
one. This knowledge can perfect *white-box* attacks, partial *black-box* attacks or
blind attacks with no adversarial knowledge. However, it is possible to attack
a DNN model F with limited knowledge in hand. In past work, such as [16],
the attacker was able to approximate the architecture of a target model, F_{target},
in a black-box setting, and create a substitute training model, which was then
used to craft adversarial examples that generalize on both models. These example were transferred back to target model, by way of *adversarial transferability*
[16] - a very powerful property, which enables an adversary to transfer malicious
examples between models to evade a target classifier model. While deep learning

networks have gathered much attention in terms of capability to solve complex and hard to solve problems, there are perilous threats that can erode and inhibit their potential [25]. It is believed that deep neural networks can be exploited from these three directions:

- *Modified Training Data* - commonly known as a *causative* or *poisoning* attack, in which the adversary influences or manipulates the training data-set χ, with a transformation. This modification could entail control over a small portion or an important determinant feature dimension D_i in the training data. With this type of attack advance, the attacker can mislead the learner in order to produce a badly classifier, which the adversary exploits post training [9].
- *Poorly Trained DNN Models* - although considered an oversight, rather than blamed on an external adversary. A perfunctory trained DNN could be due to several reasons. Most of the time, developers Credulously use DNNs prepared and trained by others. These same DNNs could have hidden vulnerabilities ripe for exploitation, which can become easy targets for manipulation by adversaries during deployment [25].
- *Perturbed Input Image* - commonly known as *adversarial examples* [12], attackers are also known to attack DNN models, during testing, by constructing malformed input to evade the learning mechanism learned of the DNN classifier. This is known as an *evasion attack* [9]. Our paper focuses on combating the this kind of attack.

A.3   Adversarial Examples

As mentioned, machine-learning models are vulnerable to adversarial attacks that seek to destabilize the neural network's ability to generalize new input; which jeopardizes the security of the model. From what we learned from the authors in [9], these attacks can either occur during the training phase as a *poisoning attack*, or testing phase as an *evasive attack*, on the classification model. In a test-time attack scenario, the attacker actively attempts to circumvent and *evade* the learning process achieved by training the model. This is done by inserting inputs that exploit *blind spots* in a poorly trained model, which cannot be easily detected. These disruptive anomalies are known as *adversarial examples*. Adversarial examples are slightly perturbed versions of regular input samples normally accepted classifiers. They are maliciously designed to have the same appearance as regular input, from a human's point of view, at least. These masquerading inputs are designed to confuse, mislead, and force the classifier to output the wrong label [8], violating the integrity of the model. These examples can be best thought of as "glitches" that can fool the deep learning model. These glitches are difficult to detect and are widely exploitable, if left unattended. To better understand them, consider this example: given an input sample x classified with function C, such that $C(x) = \ell$, producing output ℓ, that was correctly classified by model $A(\cdot)$, we say the perturbed input sample x^*, so that $C(x^*) = \ell$, we say x' is an adversarial example of x such that $A(x') = A(x)$. Classification models are considered *robust* if their classification ability is unaffected by the presence and exploits of adversarial examples. Adversarial examples x^* possess an appearance similar or *close* to the original input samples x. Normally used, although not the only form of measurement. This measure of *closeness* or *similarity* between the pair of original and modified input is known as the p-norm distance $\| x \|_p$. This degree of closeness could be l_2, which is the *Euclidean Distance* between two pixels in an input sample x, l_∞, which is the absolute or max change made to a pixel in x, or l_1; which is the total number of pixel changes made to the input sample x [3]. If the measure of distortion in any of the previous metrics of closeness is small, then those input samples must be visually similar to each other, which makes them a prime candidate for adversarial example generation.

A.4   The Adversarial Optimization Problem

Generating adversarial examples means there is a computational cost involved. In the general case, adversarial examples are generated by solving a hard optimization problem similar to the one below [18]:

$$x^* = x + argmin\{z : \hat{O}(x + z) \neq \hat{O}(x)\} = x + \delta_x$$

Where $x + \delta_x$ represents the least possible amount of *noise* added to cause a perturbation, while remaining unnoticeable by humans. The adversary wishes to produce adversarial examples x^* for a specific input sample x that will cause a misclassification by the target model T_{target}, with a queried adversarial sample, such that $O\{x + \delta_x\} = O\{x\}$. This misclassification proves that the classifier has been compromised, and is no longer usable. The misclassification error and drop in target label accuracy the attacker is after is achieved by adding the least amount of possible noise δ_x to the input x, in order to be unnoticeable by humans, but just enough to mislead the DNN. Solving for x^* is an optimization problem that is not easy to solve since it is *non-linear*, where multiple true solutions exist, and *non-convex*, where there not so easy to find. An optimization problem is considered to be *convex* if convex optimization methods can be used on the cost function $J(\theta)$, that if minimized $\min_x J_0(x)$, for the best possible and unique outcome can guarantee a global optimal solution. In convex-type problems, optimization is likely a well-defined problem here with one optimal solution or *global optimum* across all feasible search regions. On the other hand, a *non-convex* problem is one where multiple local minimums exist (solutions) exist for the cost function $J(\theta)$. Computationally, it is difficult to find one solution that satisfied all constraints. Here, optimality has become a problem, and an exponential amount of time and variables are required to find a feasible solution, where many indeed exist. By preventing the attacker from learning anything about the model T_{target} in a black-box system setting; it makes it more difficult to solve this computational challenge.

In our approach, we introduce this *difficulty* by deceiving the adversary and allowing him to attempt in solving this optimization problem, as an infeasible task for a decoy model T_{decoy}, which has no real value. Generating these adversarial examples is already exhaustive in computational cost time, as well as approximating and training the substitute decoy model to craft the examples. And if the attacker does indeed succeed in generating these examples, it would a highly infeasible task done in vanity.

A.5   Impact of Adversarial Examples on Deep Neural Nets

As it is known, a machine-learning application could be in severe jeopardy if the underlying model were to fall in the hands of an adversary, with intentions on launching an attack. However, there are certain measures taken to prevent the latter from occurring. However, equally menacing, and as likely, is if an adversary were able insert an input, image or query that would bypass the model's learning mechanism, and cause a misclassification attack, in full view of the defender. Adversarial Examples have the ability to do just that.

Deep neural nets depend on the discriminate features $X_{m,n} = (x_{1,1}, x_{1,2}, x_{1,3}, \ldots, x_{1,n})$, embedded within the image that the DNN model recognizes and learns, which it then assigns to its correct class label. However, according to [15] it was shown that the DNN models can be *tricked* and convinced that a *slightly* perturbed image or input that should otherwise be unrecognizable and consequently rejected by the neural network, can be *forced* to be generalized and accepted as a recognizable member of a class in the targeted model. The consequence of this is that many state-of-the-art machine-learning systems deployed in a real-world setting are left vulnerable to adversarial attacks, at any point in time from any user. This creates calamity, because any chosen input unrecognizable to the model can be *transformed* and classified with high confidence causing a *(false positive)*, and an input recognizable to the model can be classified with low confidence *(false negative)*, violating the integrity of a prediction model, eventually making it unusable. For instance, some of the most striking examples are in the case of audio inputs that sound unintelligible (to human), but contain voice-command instructions that could mislead the deep neural network [12]. In the case of facial recognition scenario, where the input is subtly modified with markings that a human being would recognize their identity correctly, but the model identifies them as someone else [12].

A.6   Adversarial Transferability

According to the authors in [24], the hypothesis of Adversarial Transferability is formulated as the following:

> *"If two models achieve low error for some task while also exhibiting low robustness to adversarial examples, adversarial examples crafted on one model transfer to the other."*

In simple terms, the idea behind *Adversarial Transferability* is that for an input sample x, the adversarial examples x^* generated to confuse and mislead one model m can be *transferred* and used to confuse other models $n_1, n_2, n_3, ..., n_i$, that are of homogeneous or even heterogeneous classifier architectures. This mysterious phenomena is mainly due to the determining property commonly shared by most, if not all machine-learning classifiers, which states that predictions made by these models vary smoothly around the input samples making them prime candidates for adversarial examples [8]. It is also worth noting these perturbed samples, referred to here as *adversarial examples*, do not exist in the decision space as a mere coincidence. But according to one hypothesis in [6], they occur within large regions of the classification model decision space. Here, dimensionality of the data is a crucial factor associated with the transferability of adversarial examples. The authors hypothesize that the higher dimensionality of the training data example set D, the more likely that the subspaces will intersect significantly, guaranteeing the transfer of samples between the two sub-spaces [6]. According to the above hypothesis, transferability holds true between two models *as long as both models share a similar purpose or task* [17]. Knowing this, an attacker can leverage the property of transferability to launch an preemptive attack, by training a local substitute classifier model F on sample testing data pairs (x', y'), that the chosen remote target classifier T_{target} were generalized on. Collecting these testing pairs can be formed into a training dataset $D_{training}$ of size N of similar dimensions and content. With the latter we can produce adversarial examples $\boldsymbol{x}^*$. It is also worth noting that the success rate of transferability varies depending on the type of remote target classifier the examples $\boldsymbol{x}^*$ are being transferred to. These modified examples can then be transferred to the target classifier. Hence, the same perturbations that influence model n also affect model m. Knowing that the above hypothesis is true in the general case, Papernot used this very same concept to attack learning systems using adversarial examples generated and transferred from a substitute classifier in [18], which is the same attack we also used for our designed adversary. This transfer property is an anomaly, and creates an obstacle in the face of deploying and securing machine-learning services on the cloud, enabling exploitation and ultimately attacks on black-box systems [24], as we'll see in the coming sections.

A.7   Black-Box Learning Systems

To explain a black-box threat model, we start by describing the term *black-box* system concept. A black-box is essentially a system that can be construed in terms of inputs x and outputs y, with the internal mechanisms of the system $f(x) = y$ transforming x into y remaining invisible. The functionality of the black-box can only be understood by observation, which is what the attacker depends on to begin his attack. The black-box threat model is by extension a black-box system. In our paper, we are attempting to prevent the attacker from polluting the target classifier T_{target}, by blocking transferability and access to the target model to change the prediction on the class label y. Here, we consider the adversary to be *weak* with limited knowledge, as in he can only observe the inputs inserted and outputs produced, while possessing little knowledge of the classifier itself. The adversary possesses very little, if no knowledge at all of the classifier architecture, structure, number or type of hyper-parameters, activation function, node weights, etc. Such an environment is considered to be a *black-box system* and the type of attacks are called *black-box attacks*. The adversary need not know the internal details of the system to exploit and compromise it [18].

Generally, in order to attack the model, in a black-box learning setting, the adversary attempts to generate adversarial examples, which are then transferred from the substitute classifier F to the target classifier T_{target}, in an effort to successfully distort the classification of the output labels [8]. The intention of the attacker is to train a substitute classifier in a way that is to mimic or simulate the decision space of the target classifier. For the latter purpose, the attacker continuously updates the substitute learning model and queries the target classifier (represented by the Oracle) for labels to train the substitute model, craft adversarial examples and attack the black-box target classifier.

Generally, the model being targeted is a multi-class classifier system, otherwise known as the *Oracle O*. Querying the *Oracle* represents the only capability which the attacker possesses. Querying the *Oracle O* for input x, which represents the only capability available to the attacker, as in the black-box model no access to the Oracle internal details is possible [18]. The goal of the adversary is to produce a *perturbed* version of any input x, known as an *adversarial sample* after modification, denoted x^*. This represents an attack on the integrity of the classification model (oracle) [18]. What the adversary attempts to do is solve the following optimization problem to generate the adversarial samples, as seen below:

$$x^* = x + \arg\min\{z : \hat{O}(x + z) \neq \hat{O}(x)\} = x + \delta_x$$

The adversary must able to solve this optimization problem by adding a perturbation at an appropriate rate with δ_x, to avoid human detection. The magnitude ε of the rate must be generated in such a way with the least perturbation possible in δ_x to influence the classifier, as well remain undetected by a human [18]. This is considered a hard optimization problem, since finding a minimal value to δ_x is no trivial task. Further more, removing knowledge of the architecture and training data makes it difficult to find a perturbation that satisfied the condition for successful adversarial examples secretion, where $O\{x + \delta_x\} = O\{x\}$ [18].

A.8   Transferability and Black-Box Learning Systems

Adversarial Transferability is critical for black-box Attacks, to say the least. In fact black-box systems are dependent on its success. In [25], it is suggested that the adversary can build a substitute training model F with synthetic labels S_0 collected by observing the labeling of test samples by the *Oracle O*, despite the DNN model and dataset being inaccessible. The attacker can then build a substitute model F from what he learns from O. The attacker will can then craft adversarial samples that will be misclassified by the substitute model F [16]. Now that the attacker has approximated the knowledge of the internal architecture of F, he can use it to construct. For as long as adversarial transferability holds between $F(S_0)$ and T_{target}. Adversarial examples misclassified by F will be misclassified by the target as well. In our paper, we find a way to re-channel adversarial transferability and prevent an attack. We plan to accomplish the latter via *deception*. It was Papernot in [17,18], who proposed that transferability can be used to transfer adversarial examples from one neural network to the other that share a common purpose or task, yet are dissimilar in network architecture. Transferability is essential for the success of black-box attacks on deep neural nets, which is due to the limitations imposed on the adversary, such as lack of architecture, model and training dataset knowledge. Even with limited knowledge, the adversary with the aid of the transferability property in the adversary's armaments, the adversary can train a substitute model and generate transferable examples, then transfer them to the unprepared target model, making the victim's trained model vulnerable to attack [26]. There has been much work focused on the abilities possessed by adversarial examples, and its ability to transplant itself between machine-learning techniques (DNN, CNN, SVM, etc.). Work, namely in [3,14,18], all reached the same conclusion - adversarial examples will transfer across different models trained on different dataset implementations, with different machine-learning techniques.

A.9   Honeypots

A honeypot can be thought of as a single or group of *fake* systems to collect intelligence on an adversary, by inducing him/her to attack it. A honeypot is meant to appear and respond like a real system, within a production environment. However, the data contained within the honeypot is both falsified and spurious, or better understood as *fake*. A honeypot has no real production value, instead its functionality is meant to record information on malicious activity. In the scenario that it should become compromised it contains no real data and therefore poses no threat on the production environment [13,23]. As mentioned, honeypots can be deployed with fabricated information, this can be an attractive target to outside attackers, and with the correctly engineered characteristics can be used to re-direct attackers towards decoy systems and away from critical infrastructure [7]. As mentioned above, honeypots have a wide array of enterprise applications and uses. Currently, honeypot technology has been utilized in detecting *Internet of Things* (IoT) cyberattack behaviour, by analyzing incoming network traffic traversing through IoT nodes, and gathering attack intelligence [4]. In robotics, a honeypot was built to investigate remote network attacks on robotic systems [10]. Evidently, there is an increasing need to install *red-herring* systems in place to thwart adversarial attacks before they occur, and cause damage to production systems. One of the most popular type of honeypots technologies witnessing an increase in its popularity is *High-Interaction Honeypots (HIHP)*. This type of honeypot is preferred, since it provides a real-live system for the attacker to be active in. This property is valuable, since it can potentially capture the full spectrum of attacks launched by adversaries within the system. It allows to learn as much as possible about the attacker, the strategy involved and tools used.

Gaining this knowledge allows security experts to get insight into what future attacks might look like, and better understand the current ones.

References

1. Adel Karimi: honeybits. https://github.com/0x4D31/honeybits. Accessed 27 Mar 2019
2. Akiyama, M., Yagi, T., Hariu, T., Kadobayashi, Y.: HoneyCirculator: distributing credential honeytoken for introspection of web-based attack cycle. Int. J. Inf. Secur. **17**(2), 135–151 (2018)
3. Carlini, N., Wagner, D.: Adversarial examples are not easily detected: bypassing ten detection methods. In: Proceedings of the 10th ACM Workshop on Artificial Intelligence and Security (AISec 2017), pp. 3–14 (2017)
4. Dowling, S., Schukat, M., Melvin, H.: A ZigBee honeypot to assess IoT cyberattack behaviour. In: Proceedings of the 2017 28th Irish Signals and Systems Conference (ISSC), pp. 1–6, June 2017
5. Egupov, A.A., Zareshin, S.V., Yadikin, I.M., Silnov, D.S.: Development and implementation of a Honeypot-trap. In: Proceedings of IEEE Conference of Russian Young Researchers in Electrical and Electronic Engineering, pp. 382–385 (2017)
6. Goodfellow, I.J., Shlens, J., Szegedy, C.: Explaining and harnessing adversarial examples. In: International Conference on Learning Representations 2017, p. 11 (2015)
7. Guarnizo, J.D., et al.: SIPHON: towards scalable high-interaction physical honeypots. In: Proceedings of the 3rd ACM Workshop on Cyber-Physical System Security (CPSS 2017), pp. 57–68 (2017)
8. Hosseini, H., Chen, Y., Kannan, S., Zhang, B., Poovendran, R.: Blocking transferability of adversarial examples in black-box learning systems. arXiv:1703.04318 (2017)
9. Huang, L., Joseph, A.D., Nelson, B., Rubinstein, B.I., Tygar, J.D.: Adversarial machine learning. In: Proceedings of the 4th ACM Workshop on Security and Artificial Intelligence (AISec 2011), pp. 43–58 (2011)
10. Irvene, C., Formby, D., Litchfield, S., Beyah, R.: HoneyBot: a honeypot for robotic systems. Proc. IEEE **106**(1), 61–70 (2018)
11. Kedrowitsch, A., Yao, D.D., Wang, G., Cameron, K.: A first look: using Linux containers for deceptive honeypots. In: Proceedings of the 2017 Workshop on Automated Decision Making for Active Cyber Defense (SafeConfig 2017), pp. 15–22 (2017)
12. Kurakin, A., Goodfellow, I.J., Bengio, S.: Adversarial examples in the physical world. In: International Conference on Learning Representations (ICLR), p. 14 (2017)
13. Lihet, M.A., Dadarlat, V.: How to build a honeypot system in the cloud. In: Proceedings of the 2015 14th RoEduNet International Conference - Networking in Education and Research (RoEduNet NER), pp. 190–194, September 2015
14. Liu, Y., Chen, X., Liu, C., Song, D.: Delving into transferable adversarial examples and black-box attacks. In: Proceedings of the International Conference on Learning Representations, p. 14 (2017)
15. Nguyen, A., Yosinski, J., Clune, J.: Deep neural networks are easily fooled: high confidence predictions for unrecognizable images. In: Proceedings of the IEEE Conference on Computer Vision and Pattern Recognition (CVPR), pp. 427–436 (2015)

16. Papernot, N., McDaniel, P., Wu, X., Jha, S., Swami, A.: Distillation as a defense to adversarial perturbations against deep neural networks. In: Proceedings of the 2016 IEEE Symposium on Security and Privacy (SP), pp. 582–597, May 2016

17. Papernot, N., McDaniel, P., Goodfellow, I.: Transferability in machine learning: from phenomena to black-box attacks using adversarial samples. arXiv:1605.07277 [cs], May 2016

18. Papernot, N., McDaniel, P., Goodfellow, I., Jha, S., Celik, Z.B., Swami, A.: Practical black-box attacks against machine learning. In: Proceedings of the 2017 ACM on Asia Conference on Computer and Communications Security (ASIA CCS 2017), pp. 506–519 (2017)

19. Rauti, S., Leppänen, V.: A survey on fake entities as a method to detect and monitor malicious activity. In: Proceedings of the 25th Euromicro International Conference on Parallel, Distributed and Network-based Processing (PDP), pp. 386–390 (2017)

20. Ribeiro, M., Grolinger, K., Capretz, M.A.M.: MLaaS: machine learning as a service. In: Proceedings of the 2015 IEEE 14th International Conference on Machine Learning and Applications (ICMLA), pp. 896–902, December 2015

21. Rozsa, A., Gunther, M., Boult, T.E.: Towards robust deep neural networks with BANG. In: Proceedings of the IEEE Winter Conference on Applications of Computer Vision (WACV) 2018, November 2016

22. Soule, N., Pal, P., Clark, S., Krisler, B., Macera, A.: Enabling defensive deception in distributed system environments. In: Resilience Week (RWS), pp. 73–76 (2016)

23. Suo, X., Han, X., Gao, Y.: Research on the application of honeypot technology in intrusion detection systems. In: Proceedings of the IEEE Workshop on Advanced Research and Technology in Industry Applications, pp. 1030–1032, September 2014

24. Tramèr, F., Papernot, N., Goodfellow, I., Boneh, D., McDaniel, P.: The space of transferable adversarial examples. arXiv:1704.03453 [cs, stat], April 2017

25. Xiao, Q., Li, K., Zhang, D., Xu, W.: Security risks in deep learning implementations. arXiv:1711.11008 [cs], November 2017

26. Yuan, X., He, P., Zhu, Q., Bhat, R.R., Li, X.: Adversarial examples: attacks and defenses for deep learning. arXiv:1712.07107 [cs, stat], December 2017

27. Zantedeschi, V., Nicolae, M.I., Rawat, A.: Efficient defenses against adversarial attacks. In: Proceedings of the 10th ACM Workshop on Artificial Intelligence and Security (AISec 2017), pp. 39–49 (2017)

Cloud S&P - Cloud Security and Privacy

Graphene: A Secure Cloud Communication Architecture

Abu Faisal$^{(\boxtimes)}$ and Mohammad Zulkernine

School of Computing, Queen's University, Kingston, ON, Canada
{faisal,mzulker}@cs.queensu.ca

Abstract. Due to ubiquitous-elastic computing mechanism, platform independence and sustainable architecture, cloud computing emerged as the most dominant technology. However, security threats become the most blazing issue in adopting such a diversified and innovative approach. To address some of the shortcomings of traditional security protocols (e.g., SSL/TLS), we propose a cloud communication architecture (**Graphene**) that can provide security for data-in-transit and authenticity of cloud users (CUs) and cloud service providers (CSPs). Graphene also protects the communication channel against some most common attacks such as man-in-the-middle (MITM) (including eavesdropping, sniffing, identity spoofing, data tampering), sensitive information disclosure, replay, compromised-key, repudiation and session hijacking attacks. This work also involves the designing of a novel high-performance cloud focused security protocol. This protocol efficiently utilizes the strength and speed of symmetric block encryption with Galois/Counter mode (GCM), cryptographic hash, public key cryptography and ephemeral key-exchange. It provides faster reconnection facility for supporting frequent connectivity and dealing with connection trade-offs. The security analysis of Graphene shows promising protection against the above discussed attacks. Graphene also outperforms TLSv1.3 (the latest stable version among the SSL successors) in performance and bandwidth consumption significantly and shows reasonable memory usage at the server-side.

Keywords: Cloud computing · Security protocol · Data-in-transit · Authentication · Perfect forward secrecy

1 Introduction

Security concerns such as data breaches and tampering, weak identities and access management, malicious insiders, system and application vulnerabilities and shared technology vulnerabilities have hazardous impact on the cloud as reported by the Cloud Security Alliance (CSA) [15]. To deal with these concerns, the majority cloud service providers (CSPs) implement a mixture of security and privacy controls to provide services to their customers. Cloud users

© Springer Nature Switzerland AG 2019
J. Zhou et al. (Eds.): ACNS 2019 Workshops, LNCS 11605, pp. 51–69, 2019.
https://doi.org/10.1007/978-3-030-29729-9_3

(CUs) connect to the cloud services using internet connectivity. However, existing traditional security protocols (e.g., SSL/TLS) that protect this connectivity, should be more efficient to handle cloud communication related security issues. Every now and then, a new security threat is raised. In most cases, man-in-the-middle (MITM) (including eavesdropping, sniffing, identity spoofing, data tampering), sensitive information disclosure, replay, compromised-key, repudiation and session hijacking attacks can happen in cloud communications [8, 15]. Traditional security protocols (e.g., SSL/TLS) are not always able to satisfy the growing demand of security in cloud communications for various reasons. These reasons are mainly related to maintaining middlebox compatibility, backward compatibility for older systems, downgrading due to unavailability of the selected protocol version or cipher suites and some recent attacks (e.g., BEAST, DROWN, CRIME, BREACH, WeakDH and Logjam, SSLv3 fallback, POODLE and ROBOT attacks) [3–5, 7, 10, 13, 17–19, 28].

The final draft of TLSv1.3 [30] is published recently. It claims to have some improvements over TLSv1.2 [31] in terms of security and performance. TLSv1.3 stops supporting all legacy symmetric encryption algorithms and static RSA and Diffie-Hellman cipher suites. It adds (EC)DHE to the base specification. Also, it uses only authenticated encryption with associated data (AEAD) algorithms. However, it still has some vulnerabilities. In TLSv1.3, the first two roundtrip handshake messages are merged into a single roundtrip message. This merged message includes the client key-exchange information, supported cipher suites information and "ClientHello" message together in *unencrypted* form. Most importantly, before the client receives "ServerHello" message, all communications are performed in unencrypted form. The client key-exchange information is one half of the key-exchange mechanism that is generated by random guessing of the server-side algorithm. Therefore, if the server does not agree or support that algorithm or the client sends no key-exchange information, the client needs to generate and send the key-exchange information again using the agreed algorithm which increases the roundtrip time. It still uses pre-shared key (PSK) cipher suites along with the above changes. Also, superfluous messages such as "ChangeCipherSpec" are eliminated while keeping a backdoor open for middlebox compatibility.

In this paper, we propose a comprehensive secure cloud communication architecture (**Graphene**). This architecture can effectively mitigate the existing threats of cloud communications between cloud entities. Graphene ensures security for data-in-transit and authenticity of cloud users (CUs) and cloud service providers (CSPs). It does not have any middlebox or backward compatibility. Either both parties communicate using the supported cipher suites recommended by the NIST or the secure channel cannot be established. We perform security analysis based on the man-in-the-middle (MITM) (including eavesdropping, sniffing, identity spoofing, data tampering), sensitive information disclosure, replay, compromised-key, repudiation and session hijacking attacks. Thus, we show that this architecture can efficiently mitigate these attacks. Graphene protects the cloud communication channels with significantly less negotiation and bandwidth overhead, reasonable memory usage and faster connectivity than the traditional security protocols (e.g., TLSv1.3).

Our main contribution in this paper is a comprehensive secure cloud communication architecture called Graphene. More specifically, the paper makes the following contributions:

- Graphene provides a novel high-performance cloud focused security protocol. This protocol efficiently utilizes the strength and speed of symmetric block encryption, cryptographic hash, public key cryptography and ephemeral key-exchange mechanism.
- It utilizes new highly-compact message structures to support secure session establishment, reconnection and data transmission. These message structures help achieve minimal bandwidth consumption and reasonable memory usage than TLSv1.3 (the latest stable version among the SSL successors) and embed other communication protocols inside it.
- Graphene ensures security of the data-in-transit and all associated secret keys. It maintains perfect forward secrecy (PFS) by performing ephemeral key-exchange on each session and encrypting the session with a new secret key.
- It is applicable to both TCP and UDP-based communications. It has no dependency on the SSL/TLS/DTLS implementations at any part of the communication channel.

The rest of this paper is organized as follows. Section 2 presents the related work. Section 3 discusses the proposed secure cloud communication architecture. Section 4 provides the implementation and experimental environment of the architecture. Section 5 presents the results. Finally, Sect. 6 summarizes the paper.

2   Related Work

The main purpose of the existing cloud security research is to secure data during cloud communications (data-in-transit) as well as the cloud data storages (data-at-rest).

Google [20–22] uses multi-layer encryption model to secure the data-at-rest while relying on default TLS for protecting the data-in-transit. Amazon Web Services (AWS) [9] and Microsoft Azure [27] focus on protecting data integrity using keyed-HMAC. AWS uses temporary non-stored session keys in EC2 load balancers and Azure uses two-factor authentication to prevent unauthorized access of data. However, they still use secure sockets layer (SSL) to provide transmission protection to their customers. They also maintain MD5 compatibility for older systems. It is clearly visible that the CSPs are mostly concerned to secure the data stored in their data centers by using multi-layer encryption model, keyed-HMAC, two-factor authentication etc. However, they rely on default TLS and sometimes even SSL for protecting the data-in-transit which makes the existing implementations vulnerable to all recent SSL/TLS related attacks [3–5,7,10,13,17–19,28]. On the contrary, Graphene ensures security of data-in-transit and authenticity of cloud entities. It does not support any security techniques which have any known vulnerabilities. Graphene has its own novel protocol, highly compact message structures and secure session management. It provides higher level of security with lower level of bandwidth consumption and reasonable memory usage.

AbdAllah et al. [6] propose a generic trust model (TRUST-CAP) for cloud-based applications by focusing on infrastructure-as-a-service (IaaS). However, it does not provide any specific protocol for securing cloud communications. Conversely, Graphene provides a cloud-focused security protocol. It ensures adequate protection to the communication channel and its associated cloud entities against some most common cloud attacks. As Graphene follows the security objectives of the TRUST-CAP model, it can be used as the security protocol for cloud communications in TRUST-CAP as well.

Kaaniche et al. [23] propose a cloud data sharing framework (CloudaSec) that encrypts the data at the server-side using the hash of the data as the symmetric key. Then, it encrypts the symmetric key using recipient's public key and includes that in the response metadata. Basically, it uses a form of hybrid cryptography [2] where the symmetric key remains the same for unchanged data. On the other hand, Liang et al. [26] and Chandu et al. [14] also propose a similar approach that follows the generic hybrid cryptography. It uses AES to encrypt the data at the client-side. Then, it encrypts the AES key using owner's RSA public key. After that, it uploads both the encrypted data and the encrypted key to the cloud storage. Both approaches are vulnerable to compromised-key, permanent data tampering, identity spoofing, MITM and MATE attacks.

Khanezaei et al. [24] propose a secure cloud storage service using server's public-key to perform RSA encryption to protect the data during communication and storage. The service generates an AES secret key and stores it in the database along with the RSA encrypted data for future sharing. However, this solution imposes tremendous computing overhead on the cloud server for decrypting large amount of RSA encrypted data on every request which is a very cumbersome and slow process. The RSA encrypted data uses server's public key which can be decrypted easily using server's private key. Moreover, the AES secret key is stored along with the RSA encrypted data which makes this solution highly prone to compromised-key, permanent data tampering, identity spoofing, MITM and MATE attacks.

Kerberos [29] is a network authentication protocol that provides authentication in a non-secured environment. It uses a key distribution center (KDC) which receives request for tickets from the clients. Then, the KDC generates ticket-granting tickets (TGT) and encrypts it using client's password. After receiving the encrypted response, the client decrypts it using the password. This is different from the central key server (CKS) mechanism in Graphene. The CKS is a root public key management system which helps Graphene architecture to prevent MITM in all phases. It is designed to store, revoke and distribute root public keys securely.

All the above discussed related work are focused on a specific facet of security and operational behavior. They are mostly concerned about the data-at-rest and their confidentiality, integrity or access control. Also, some recent attacks such as BEAST, DROWN, CRIME, BREACH, WeakDH and Logjam, POODLE, ShellShock and ROBOT attacks [3–5,7,10,13,17–19,28] have shown that the existing security protocols are not able to mitigate the increasing threats of cloud communications. It becomes a major hindrance while expanding towards IoT, fog or edge computing, connected vehicles etc. Therefore, a comprehensive secure cloud communication architecture is mandatory to mitigate the rising threats against cloud communications.

3   Graphene Architecture

This section presents the proposed secure cloud communication architecture in detail. In the following section, we discuss about the design of this architecture and different communication phases of it. In Sect. 3.2, we explain the sequence of events executed at both user and server ends.

3.1   Design Specification

Graphene focuses on the security of data-in-transit in cloud computing. It guarantees the authenticity of cloud entities by using a new Central Key Server (CKS) mechanism. The CKS is designed to store, revoke and distribute root public keys securely. Graphene efficiently combines and utilizes the strength and speed of the symmetric block encryption, cryptographic hash, public key cryptography and ephemeral key-exchange mechanism. Symmetric encryption provides confidentiality, cryptographic hash enables integrity and public key cryptography ensures authenticity and non-repudiation. It embeds these four essential security elements into the communications in such a way so that a cloud user (CU) can have a secure communication channel with the Cloud Front End (CFE) server. It ensures security for both the data and the cryptographic keys.

The system does not use any long-term keys. Each session is encrypted with a new secret key thus ensuring perfect forward secrecy (PFS). Graphene is applicable for both TCP and UDP-based communications. It works in the application layer. Thus, it can be easily integrated with any protocols and server systems. Graphene utilizes seven highly compact new message structures: (i) publish (PUB), (ii) acknowledge (ACK), (iii) reconnect (RECON), (iv) request (REQ), (v) response (RES), (vi) expired (EXP), and (vii) error (ERR). They make Graphene more efficient in terms of performance, bandwidth consumption, memory usage and integration with the existing protocols. These message structures facilitate the secure session establishment, reconnection, data transmission and error handling between cloud entities.

The architecture consists of six different communication phases such as registration, initialization, session establishment, data transmission, termination and reconnection. First, the cloud entities need to register their root public keys to

the central key server (CKS) in the registration phase. After that, when any cloud
user wants to communicate to the cloud server, temporary cryptographic keypairs and hash functions are initialized in the initialization phase to establish an
encrypted session. Then, both the entities exchange their temporary public keys
with each other, signed by their respective root private key. The key-exchange
of temporary public keys is secured by hybrid-crypto mechanism [2,16] using
AES-GCM for data encryption and RSA/ECC for key encryption during the
session establishment phase.

After that, both entities generate common symmetric encryption key using
ephemeral key-exchange. Then, they start transmitting encrypted signed data
to each other in the data transmission phase. After sending the response payload
successfully to the cloud user, cloud server terminates the connection which is
called the termination phase. At this point, the server keeps the encrypted session
information till the session expires. Within that period, the cloud user can send
a reconnection request and re-establish the encrypted session for further data
transmission which is called the reconnection phase. Figure 1 shows these phases
of communications used by Graphene which we discuss in detail in the following
paragraphs.

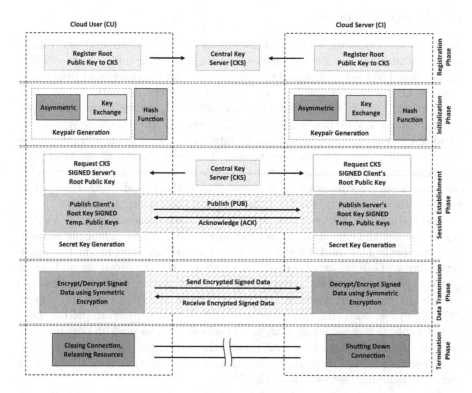

Fig. 1. Different communication phases between cloud entities in Graphene
architecture

Registration Phase. All cloud entities must register their root public keys to the central key server (CKS) prior to any communication. The CKS public keys must also be installed in the cloud entity systems, to ensure integrity and authenticity of the data communicated between the CKS and the cloud entities. The CKS itself and all communications (key registration, revocation and distribution) with it are assumed to be secured at this point.

Initialization Phase. In case of cloud server instance (CI), this phase occurs at the very beginning when the CI is initiated. However, for the cloud user (CU), it occurs when a new cloud connection is created to commence communication with the cloud front end (CFE) server. During this phase, each cloud entity generates a pair of temporary public-private keypairs. One keypair (RSA/ECC) is for maintaining the authenticity and integrity of the payloads. The other keypair (DHE/ECDHE) is for the ephemeral key-exchange. Each cloud entity also initializes cryptographic hash functions according to the design specification.

Session Establishment Phase. When a CU tries to connect to the CI for the first time, a temporary encrypted session is initialized between the CU and the CI. During this time, a pair of messages (PUB-ACK) are transmitted between them. Both parties store the other party's pair of public keys in that temporary session protected by a 64-byte hashed session key. Then, they generate a common secret key to proceed with the data transmission phase. The 64-byte hashed session key is updated after every successful transaction (request-response). The CU always receives the updated session key hidden inside the encrypted response. When this session expires, all the negotiated public keys and generated common secret key are destroyed automatically.

Data Transmission Phase. After establishing the secure session, both parties use the common secret key to perform symmetric block encryption for maintaining the confidentiality of the request and response payloads. The negotiated temporary keypair (RSA/ECC) is used to perform payload signing and verification that ensures authenticity and integrity of the payload throughout the session. Every signing operation performed in this architecture involves timestamp to protect against replay attacks. During this phase, a cloud focused cryptographic hash function (Blake2b [1]) is used to protect the data integrity.

Termination Phase. In this phase, when the CI sends encrypted response back to the CU successfully, the communication channel is terminated. The existing session remains valid for reconnection until it is expired.

Reconnection Phase. This phase is not explicitly shown in Fig. 1. It has implicit activity in this architecture. After the termination phase, if the CU again connects to the server and sends a valid reconnection (RECON) packet with the last received session key, the encrypted session is re-established between the CU and the CI. The CFE maintains a session key mapping of the CIs. Based on the session key, it reconnects the CU to the appropriate CI. Both parties use the previously negotiated pair of public keys and the stored common secret key. Therefore, re-keying the block cipher during the session is not needed.

3.2 Flow of Execution

This section explains how this architecture establishes a secure encrypted channel for communications and all the internal steps illustrated by the sequence diagram shown in Fig. 2.

Step-1. In this step, the cloud user (CU) initializes a cloud connection. A pair of temporary public-private keypair is generated and the cryptographic hash functions are initialized.

Step-2. After initializing the connection, the cloud user fetches the cloud server's root public key which is signed by the central key server (CKS) that ensures authenticity and non-repudiation for both parties.

Step-3. The cloud user (CU) connects to the cloud front end (CFE) server and a cloud instance (CI) is allocated for this connection.

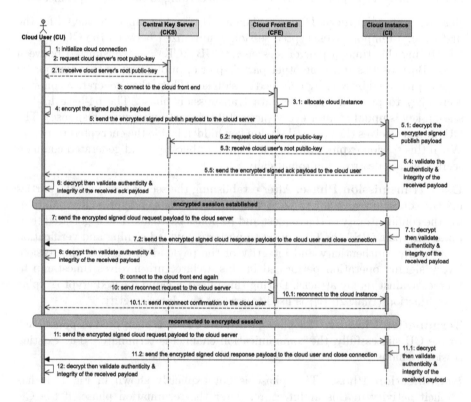

Fig. 2. Sequence diagram showing the flow of execution in Graphene architecture

Step-4. The CU signs its temporary public keys with own root private key to protect authenticity and integrity of the "publish" payload (PUB). After that, the signed "publish" payload is encrypted using symmetric block encryption to maintain the confidentiality of the payload in a hybrid-crypto mechanism [2,16].

Step-5. The CU and CI utilize the "publish" and "acknowledge" packets (PUB-ACK) to share all temporarily generated public keys to each other. The CU sends the encrypted signed "publish" payload to the CI. After decrypting the received packet, the CI requests the cloud user's root public key from the central key server (CKS). Then, the CI validates the authenticity and integrity of the received "publish" payload. After validation, the CI sends the encrypted signed "acknowledge" payload (ACK) to the CU. This approach protects the session establishment phase from man-in-the-middle (MITM) attacks.

Step-6. When the CU receives the encrypted signed "acknowledge" packet (ACK), it also validates the authenticity and integrity of the received payload. The cloud user stores cloud server's temporary public keys in the session. After finishing session establishment phase, the common secret key is generated at both ends using the ephemeral key-exchange mechanism (DHE or ECDHE). A secure encrypted communication channel is established without using any pre-shared key or transmitting any part of the secret key. This generated secret key is used to perform symmetric block encryption on the signed cloud payload.

Step-7, 8. In this step, both parties perform data transmission (request-response) which is first signed and then encrypted to protect confidentiality, integrity and authenticity of the data. After sending the response, the cloud instance (CI) terminates the connection with the cloud user (CU).

Step-9, 10. When the CU again wants to accomplish any more data connectivity and it has the valid session information, it can send a reconnection packet (RECON) to the cloud front end (CFE) server. If any associated session is found, the secure channel is re-established between the CU and the CI. They do not need to perform the session establishment steps again. Otherwise, the CU must go through Step-4 to Step-6 again.

Step-11, 12. Once the secure session is re-established, both the CU and the CI can do data transmission again. After the response is sent back to the CU, the CI closes the connection.

4 Implementation and Experimental Environment

This section explains the implementation and experimental environment used to evaluate Graphene in terms of performance, bandwidth consumption and memory usage. In the following section, we briefly discuss the implementation details of the architecture. Then, in Sect. 4.2, we explain the experimental environment.

4.1 Implementation

Graphene is developed using Java and Java Cryptography Architecture (JCA). It has no dependency on any other platforms, tools and libraries. Therefore, our implementation can be deployed in any platform or environment where Java runtime environment (JRE) is available. To compare Graphene against TLSv1.3, we run all our experiments in Java11.0.1 (LTS) which includes an implementation of the TLSv1.3 specification [30]. A novel high-performance cloud focused security protocol is designed and implemented with seven highly compact message structures. Any types of payload data (e.g., HTTP, XML, JSON and Binary) can be sent and received using this protocol with minimal changes in the existing infrastructures and applications.

Graphene uses public-key cryptography for signing the payloads and ephemeral Diffie-Hellman (at least 2048-bit) using MODP groups [25] as the key-exchange mechanism. A latest cryptographic hashing algorithm Blake2b [1] is used for maintaining the integrity of the data-in-transit. It is faster than SHA-families and as secure as SHA-3 at minimum, which makes it a perfect candidate for cloud communications and large volume of data hashing. SHA-512 is used to generate temporary session keys from the connection properties and the client supplied information. AES-256 with Galois/Counter mode (GCM) is used as the symmetric block encryption for ensuring confidentiality throughout all the communication phases. The system operates over 256-bit encrypted channel which is the approved encryption standard for *top secret* information by both the National Institute of Standards and Technology (NIST) and the National Security Agency (NSA) of the USA.

This architecture is configurable to use any of the supported (RSA/ECC) public-key cryptographic algorithms for payload signing and verification. However, the minimum key size recommended by the NIST is 2048-bit for RSA and 224-bit for ECC [12]. Our implementation *strictly* follows these recommendations made by the NIST at all steps [11,12]. AES (128/192/256-bit) encryption is used as the supported symmetric block encryption in Graphene. AES-256 is the highest level (military-grade) of symmetric encryption available at present. It is also the default choice for confidentiality in Graphene. However, Graphene can be configured to use any of the other key sizes or encryption algorithms if this level of security is not required.

4.2   Experimental Environment

As illustrated in Fig. 3, cloud instances (CIs) are configured according to the requirement. Each CI has 1 hyper-threaded vCPU core (4.0 GHz frequency with turbo boost), 4 GB of RAM, 20 GB of local SSD storage. Each cloud instance runs CentOS 7 (*minimal version*) to have less interference from other processes. The cloud instances are setup and controlled by a cloud front end (CFE) server. The CFE server has a built-in basic load balancer which works in a simplified round-robin fashion. It is responsible for distributing all incoming traffics to these cloud instances equally by assigning the same weight to each instance (CI) unless the incoming traffic is a reconnection request with valid session information.

The CI records execution time for session establishment (if any), request and response at the server-side for plaintext, TLSv1.3, TLSv1.2 and Graphene with and without session-reconnection mechanism. However, the cloud user (CU) monitors roundtrip time information at the client-side for further analysis. All CUs run in an iterative fashion and send request with a specific size (100B, 500B,

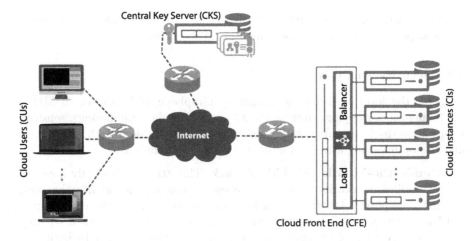

Fig. 3. Experimental environment of the Graphene architecture

1 KB, 500 KB or 1 MB) of data every time. A separate secure public key registration and distribution server runs as a central key server (CKS) for managing root public keys. In CKS, all cloud entities have their root public keys registered against their unique identifier. In Graphene, the CFE server and the CUs have their public keys registered against their IP addresses and assigned random string tokens. All experiments are performed in an iterative fashion (1000 times). Each request belongs to a temporary encrypted session which has a hashed session key generated from the connection properties and the client supplied information.

The reason behind comparing with TLSv1.3 in our experiment is that it is the latest stable version among the SSL (Secure Sockets Layer) successors. TLSv1.3 is claimed to be more secure than TLSv1.2, where TLSv1.2 is proved to have a steady and secure implementation than SSL, TLSv1.0 and TLSv1.1. SSLv3 and TLSv1.0 are already declared obsolete and some vulnerabilities are reported for TLSv1.1. Due to the severe data breaches caused by recent attacks, TLSv1.3 is now recommended for secure communications over the internet. If TLSv1.3 is not available, at least TLSv1.2 should be used for secure communications.

5 Results and Analysis

This section presents the results and analyzes the solution. All prominent cryptographic technologies (public key cryptography, digital signature and verification, symmetric block encryption and cryptographic hash) are evaluated iteratively for different payload sizes (100B–20 MB) to select the optimal choice for implementing a high-performance cloud focused security protocol (i.e. Graphene) that efficiently utilizes these technologies with respect to their strength and speed. The following section presents a thorough security analysis of Graphene against different types of attacks. After that, we evaluate the performance of Graphene in terms of execution time on server-side, roundtrip time on client-side, bandwidth overhead with respect to plaintext, memory usage at server-side and impact of different payload sizes in the above mentioned scenarios.

5.1 Security Analysis

To show the level of defense provided by Graphene with respect to MITM (including eavesdropping, sniffing, identity spoofing, data tampering), sensitive information disclosure, replay, forward secrecy (compromised-key), repudiation and session hijacking attacks, we provide a thorough analysis.

(i) Man-in-the-Middle (MITM) Attack. This attack is basically a combination of different security attacks like eavesdropping, sniffing, identity spoofing and data tampering. In MITM attacks, an adversary can actively eavesdrop to a private communication between two legitimate users or even create separate connections to each of the users to appear as a legitimate entity to both parties (identity spoofing). Then, the attacker captures all the packets (sniffing) and forwards them to the other party in such a way so that the victims are forced to believe that they are communicating directly to each other over a private connection. In the later approach, the attacker has full control over the communication and can easily steal valuable information or even manipulate the packets (data tampering) sent to the victims. In order to analyze Graphene against these attacks, we investigate two types of connections made from any entity in Graphene. One is from cloud user (CU) or cloud instance (CI) to the central key server (CKS) and the other is in between CU and CI as discussed below.

(a) CU/CI to CKS. When any CU/CI requests any public key from the CKS, the CKS responds with the requested public key payload signed by its own root private key. The root public key of CKS is installed to all entity systems during setup time. Thus, the receiver can verify the authenticity and integrity of the received public key payload from the CKS which prevents identity spoofing and data tampering. Since the payload is a public key and it is meant to be shared publicly, confidentiality of this type of payload is not required at all. Therefore, even if any adversary is eavesdropping or sniffing to this connection, the adversary cannot tamper with the payload. Hence, MITM attacks are not possible for this type of connection.

(b) Between CU and CI. All communications between the CU and the CI are securely protected (signed and encrypted). Each packet is signed by their root or temporary private key based on the communication phase. Thus, the other entity can always verify the authenticity of the sender by using sender's root or temporary public key. Signing each packet ensures the authenticity and integrity of the received payload in all phases which prevents the identity spoofing and data tampering attacks on DHE key-exchange and request-response payloads. Finally, due to AES-GCM encryption, the adversary can never see the payloads transmitted through this channel at any time which eliminates the scope of eavesdropping or sniffing. Thus, ensuring MITM attacks cannot be successful on this connection at all.

(ii) Sensitive Information Disclosure. This attack often happens where the payload is transmitted in plaintext or the encryption technique used is prone to cryptanalysis attacks. In this scenario, the adversary can capture all the packets and steal transmitted sensitive information without the knowledge of the user. However, in Graphene, all communications between CU and CI are performed using AES-GCM encrypted channel (at least 128-bit) from the transmission of first packet. Thus, no sensitive information can be accessed without establishing a proper communication channel.

(iii) Replay Attack. This is one of the most common attacks which helps the attacker to intercept valid payloads and retransmit those captured payloads repeatedly to perform some malicious or fraudulent activities. In Graphene, we designed the architecture in a manner so that this kind of attack cannot be successful. First, all our payload signing involve timestamp to create randomness in the output. Then, temporary session key is updated after every successful transaction (request-response) during the data transmission phase. This timestamp-based signing and temporary session key enable Graphene to prevent replay attacks. Thus, at no point, an adversary can gain any benefit from repeating any previously captured data.

(iv) Forward Secrecy. In cryptography, forward secrecy is a feature that ensures compromising any secret key does not compromise the security of the past payloads communicated between the entities. In our approach, we maintain perfect forward secrecy (PFS) through ephemeral Diffie-Hellman key-exchange with at least 2048-bit key size on each new session and by generating all associated cryptographic keys per session as well. Therefore, even if one session is compromised, other past and future sessions remain secure.

(v) Repudiation. This means denying the responsibility of any actions performed. In Graphene, all entities must be registered to CKS prior to any communication. The session establishment phase is performed using their registered root public-private keypair and both entities (CU and CI) negotiate temporary keypairs for this session. Later on, all communications are authenticated using these temporary public-private keypairs. This ensures authenticity and non-repudiation of the entities throughout this communication. Thus, this attack is not feasible by any means over this communication channel.

(vi) Session Hijacking. In session-based communications, attackers often try to capture session related information. More specifically, they try to lookup session keys or nonce information. In our approach, we use temporary hashed session keys generated based on connection properties and client supplied information. This session key enables cloud entities to re-establish their previous encrypted session if not expired already. Each session key is updated after every successful transaction (request-response) and most importantly, all transmitted packets in Graphene are AES-GCM encrypted.

(vii) Some Recent Attacks. Some hazardous attacks such as DROWN, CRIME, BREACH, BEAST, WeakDH and Logjam, SSLv3 fallback, POODLE and ROBOT attacks [3–5,7,10,13,17–19,28] happen on traditional security protocols (e.g., SSL/TLS) that highly threaten the existing cloud infrastructures and their expansion towards fog or edge computing, IoT, connected vehicles, smart city etc. Some of the attacks are performed by exploiting weaknesses in the security technologies whereas some are caused by misconfiguration of the system. Due to the advancement of computing resources, security measures which deemed secure in the past become vulnerable to brute force attacks, adaptive chosen plaintext attacks, compression ratio leak, discrete logarithm or other cryptanalysis attack techniques. Graphene strongly follows the NIST recommendations in choosing suitable cryptographic algorithms and their minimum supported key sizes. This enables Graphene to prevent such attacks. It uses Galois/Counter mode (GCM) as the mode of operation for AES with new initialization vector (IV) values for each request. Graphene does not deal with any compression techniques. It strictly follows the recommended key sizes by the NIST [11,12] for the minimum level of security and also uses MODP [25] groups (group id 14 or above) to perform ephemeral key-exchange.

5.2   Performance Analysis

This section presents the performance evaluation of the implemented architecture in terms of average execution time on the server-side, roundtrip time on the client-side, bandwidth overhead with respect to plaintext, TLSv1.3 and TLSv1.2 communications and memory usage at the server-side. Table 1 represents the specification of the experimental environment used for evaluating performance, bandwidth overhead and memory usage.

Table 1. Cloud Instance Specification

Parameters	Values
Virtual CPU(s), Memory	vCPUs: 1 (HyperThreaded), RAM: 4 GB
VM Class	Regular (Non-Preemptible)
Processing Unit	4.0 GHz with Turbo Boost (8M Cache)
Cloud OS & Storage	CentOS 7 (Minimal) with 20 GB SSD Storage
CFE Load Balancer	Round Robin
Sample Data	100B, 500B, 1 KB, 500 KB, 1 MB
Number of Iteration	1000

Figure 4(a) shows the average execution time for one of the investigated cloud instances in milliseconds. We investigate the average execution times in different cloud instances for plaintext (yellow curve), TLSv1.3 (purple curve), TLSv1.2 (orange curve), Graphene without session-reconnection (blue curve) and Graphene with session-reconnection (green curve) for different payload sizes (100B, 500B, 1 KB, 500 KB and 1 MB).

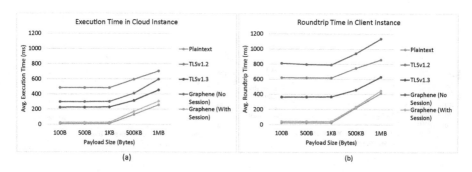

Fig. 4. Comparison of average (a) server-side execution time and (b) client-side roundtrip time in Graphene architecture (with/without session-reconnection) with respect to plaintext, TLSv1.3 and TLSv1.2 communications for different payload sizes (Color figure online)

Graphene with session-reconnection mechanism (green curve) outperforms TLSv1.3 (purple curve) significantly for all payload sizes and lies very close to the plaintext (yellow) curve and behaves the same in all cloud server instances. Graphene with session-reconnection (green curve) performs around 90% faster than the TLSv1.3 communication. Our solution even shows better results with session-reconnection (green curve) and without session-reconnection (blue curve) mechanism with respect to TLSv1.2 (orange curve).

On the client-side, we have measured the average roundtrip time (in milliseconds) by taking the sum of observed durations for connection creation, session establishment (if present) and request-response time for different payload sizes. Figure 4(b) presents the average roundtrip time for one of the investigated client instances under plaintext (yellow curve), TLSv1.3 (purple curve), TLSv1.2 (orange curve), Graphene without session-reconnection (blue curve) and Graphene with session-reconnection (green curve) for different payload sizes (100B, 500B, 1KB, 500KB, 1MB).

As observed from the performance curves of client-side average roundtrip time, Graphene with session-reconnection mechanism (green curve) performs very close to that of the plaintext (yellow) curve and shows promising performance against TLSv1.3 (purple curve). The performance of Graphene without session-reconnection mechanism (blue curve) deteriorates in terms of average roundtrip time at the client-side. However, if it is used with session-reconnection mechanism, it is able to provide faster communication with higher level of security.

The bandwidth overhead graph shown in Fig. 5(a) is calculated with respect to the bandwidth consumption of the plaintext communication. It is readily noticed that the bandwidth overhead for 100 bytes of payload size is more than 280% for TLSv1.3 (purple column) and over 380% more for Graphene without session-reconnection mechanism (blue column). However, when Graphene is used with session-reconnection mechanism (green column), it shows only 80% overhead with respect to plaintext communication and provides 54% gain over TLSv1.3 communication.

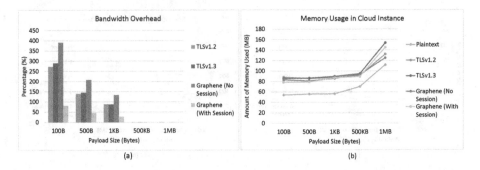

Fig. 5. Comparison of (a) bandwidth overhead and (b) average server-side memory usage in Graphene architecture (with/without session-reconnection) with respect to plaintext, TLSv1.3 and TLSv1.2 communications for different payload sizes (Color figure online)

For 1 KB of payload size, Graphene with session-reconnection mechanism provides 32% gain over the bandwidth consumption of TLSv1.3. The graph shows a decreasing trend with increasing payload sizes and for 500 KB payload size the overhead becomes nearly 1% for all types of communications with respect to plaintext. Therefore, in case of large volume of data, it seems like the overhead is negligible. However, Graphene with session-reconnection performs noticeably well in smaller payload sizes as well as with the increasing payload sizes.

Figure 5(b) shows the server-side memory usage (in MB) of Graphene in one of the investigated cloud instances with respect to plaintext, TLSv1.3 and TLSv1.2 communications. From the figure, it is readily noticed that Graphene with and without session-reconnection mechanism shows reasonable amount of memory usage for different payload sizes which lies very close to the memory usage of TLSv1.3 and TLSv1.2 communications. The usage pattern shows similar behavior in all the investigated cloud instances and the memory usage increases proportionally with the increase in payload size.

Overall, Graphene with session-reconnection mechanism performs significantly better than the TLSv1.3 in terms of server-side performance, client-side roundtrip time, bandwidth overhead and memory usage at server-side. Once the session establishment phase is complete, it can efficiently establish 256-bit encrypted channel without causing any performance, bandwidth or memory overhead. However, Graphene without session-reconnection mechanism performs worse than TLSv1.3 because of the temporary keypair generations in each session at both ends (client and server). In every session, two temporary keypairs are generated at each side to establish the session. Communicating with the central key server (CKS) by the cloud user and the cloud instance does not have that much impact on the roundtrip time. Also, Graphene was not evaluated against TLSv1.3 0-RTT mode due to unavailability of the implementation of this mode in Java11.0.1 (LTS).

6 Conclusion

Most recent security attacks and vulnerabilities of the traditional security protocols (SSL/TLS), are the major road blocks in the expansion of cloud computing. In this paper, we propose a comprehensive secure cloud communication architecture (Graphene) that mitigates these attacks and vulnerabilities. In Graphene, security of data-in-transit and authenticity of cloud entities are ensured and firmly integrated into the communications to protect against wide range of cloud attacks. A novel high-performance cloud focused security protocol is designed and implemented. It has seven highly compact new message structures which establish a secure performance and bandwidth-efficient protocol with reasonable memory usage. This architecture can successfully prevent man-in-the-middle (MITM) (including eavesdropping, sniffing, identity spoofing, data tampering), sensitive information disclosure, replay, compromised-key, repudiation and session hijacking attacks. Graphene with session-reconnection mechanism shows 90% faster execution time than TLSv1.3 (the latest stable version among the SSL successors) on the server-side and exhibits similar performance at the client-side as well. In terms of bandwidth consumption, it shows 54% gain over TLSv1.3 and overall reasonable memory usage against different payload sizes. It enforces the NIST recommendation as the base level of security for data-in-transit in cloud computing. In the future, we will work on the applications of this architecture in different sectors.

Acknowledgment. This work is partially supported by the Natural Sciences and Engineering Research Council of Canada (NSERC) and the Canada Research Chairs (CRC) program. We would also like to convey special thanks to Mohima Hossain from the TRL Lab at Queen's University for the fruitful discussion and her critics during this research work.

References

1. BLAKE2 - fast secure hashing (2017). https://blake2.net/. Accessed 02 Sept 2018
2. Hybrid CryptoSystem (2017). https://en.wikipedia.org/wiki/Hybrid_cryptosystem. Accessed 02 Sept 2018
3. Weak Diffie-Hellman and the Logjam Attack (2017). https://weakdh.org/. Accessed 02 Sept 2018
4. CRIME (2018). https://en.wikipedia.org/wiki/CRIME. Accessed 02 Sept 2018
5. Transport Layer Security: Attacks against TLS/SSL (2018). https://en.wikipedia.org/wiki/Transport_Layer_Security#Attacks_against_TLS/SSL. Accessed 02 Sept 2018
6. Abdallah, E.G., Zulkernine, M., Gu, Y.X., Liem, C.: Trust-cap: a trust model for cloud-based applications. In: 2017 IEEE 41st Annual Computer Software and Applications Conference (COMPSAC), vol. 2, pp. 584–589, July 2017. https://doi.org/10.1109/COMPSAC.2017.256
7. Adrian, D., et al.: Imperfect forward secrecy: how Diffie-Hellman fails in practice. In: Proceedings of the 22nd ACM SIGSAC Conference on Computer and Communications Security, CCS 2015, pp. 5–17. ACM, New York (2015). https://doi.org/10.1145/2810103.2813707

8. Amara, N., Zhiqui, H., Ali, A.: Cloud computing security threats and attacks with their mitigation techniques. In: 2017 International Conference on Cyber-Enabled Distributed Computing and Knowledge Discovery (CyberC), pp. 244–251, October 2017. https://doi.org/10.1109/CyberC.2017.37

9. Amazon Web Services: Amazon Web Services: Overview of Security Processes, May 2017. https://d1.awsstatic.com/whitepapers/Security/AWS_Security_Whitepaper. pdf. Accessed 02 Sept 2018

10. Aviram, N., et al.: Drown: breaking TLS using SSLv2. In: USENIX Security Symposium, pp. 689–706 (2016)

11. Barker, E.B., Dang, Q.H.: SP 800-57 Pt3 R1. Recommendation for key management, part 3: application-specific key management guidance, January 2015. https://nvlpubs.nist.gov/nistpubs/SpecialPublications/NIST.SP.800-57Pt3r1.pdf. Accessed 02 Sept 2018

12. Barker, E.B., Roginsky, A.L.: SP 800-131A R1. Transitions: recommendation for transitioning the use of cryptographic algorithms and key lengths, November 2015. http://nvlpubs.nist.gov/nistpubs/SpecialPublications/NIST.SP. 800-131Ar1.pdf. Accessed 02 Sept 2018

13. Böck, H., Somorovsky, J., Young, C.: Return of bleichenbacher's oracle threat (ROBOT). In: Proceedings of the 27th USENIX Conference on Security Symposium, SEC 2018, pp. 817–832. USENIX Association, Berkeley (2018). http://dl. acm.org/citation.cfm?id=3277203.3277265. Accessed 02 Sept 2018

14. Chandu, Y., Kumar, K.S.R., Prabhukhanolkar, N.V., Anish, A.N., Rawal, S.: Design and implementation of hybrid encryption for security of IoT data. In: 2017 International Conference On Smart Technologies For Smart Nation (SmartTechCon), pp. 1228–1231, August 2017. https://doi.org/10.1109/SmartTechCon.2017. 8358562

15. Cloud Security Aliance: the treacherous 12 - top threats to cloud computing + industry insights, October 2017. https://cloudsecurityalliance.org/download/ artifacts/top-threats-cloud-computing-plus-industry-insights/. Accessed 02 Sept 2018

16. Cramer, R., Shoup, V.: Design and analysis of practical public-key encryption schemes secure against adaptive chosen ciphertext attack. SIAM J. Comput. **33**(1), 167–226 (2004). https://doi.org/10.1137/S0097539702403773

17. Duong, T., Rizzo, J.: Here come the XOR ninjas. White paper, Netifera (2011)

18. Durumeric, Z., et al.: The matter of heartbleed. In: Proceedings of the 2014 Conference on Internet Measurement Conference, IMC 2014, pp. 475–488. ACM, New York (2014). https://doi.org/10.1145/2663716.2663755

19. Fardan, N.J.A., Paterson, K.G.: Lucky thirteen: breaking the TLS and DTLS record protocols. In: 2013 IEEE Symposium on Security and Privacy, pp. 526–540, May 2013. https://doi.org/10.1109/SP.2013.42

20. Google: Encryption at Rest in Google Cloud Platform, August 2016. https://cloud. google.com/security/encryption-at-rest/default-encryption/resources/encryption-whitepaper.pdf. Accessed 02 Sept 2018

21. Google: Encryption in Transit in Google Cloud, November 2017. https:// cloud.google.com/security/encryption-in-transit/resources/encryption-in-transit-whitepaper.pdf. Accessed 02 Sept 2018

22. Google: Google Infrastructure Security Design Overview, January 2017. https:// cloud.google.com/security/infrastructure/design/resources/google_infrastructure_whitepaper_fa.pdf. Accessed 02 Sept 2018

23. Kaaniche, N., Laurent, M., Barbori, M.E.: CloudaSec: a novel public-key based framework to handle data sharing security in clouds. In: 2014 11th International Conference on Security and Cryptography (SECRYPT), pp. 1–14, August 2014
24. Khanezaei, N., Hanapi, Z.M.: A framework based on RSA and AES encryption algorithms for cloud computing services. In: 2014 IEEE Conference on Systems, Process and Control (ICSPC 2014), pp. 58–62, December 2014. https://doi.org/10.1109/SPC.2014.7086230
25. Kivinen, T., Kojo, M.: More modular exponential (MODP) Diffie-Hellman groups for internet key exchange (IKE) (2003). https://tools.ietf.org/html/rfc3526. Accessed 02 Sept 2018
26. Liang, C., Ye, N., Malekian, R., Wang, R.: The hybrid encryption algorithm of lightweight data in cloud storage. In: 2016 2nd International Symposium on Agent, Multi-Agent Systems and Robotics (ISAMSR), pp. 160–166, August 2016. https://doi.org/10.1109/ISAMSR.2016.7810021
27. Microsoft: Trusted Cloud: Microsoft Azure Security, Privacy and Compliance, April 2015. http://download.microsoft.com/download/1/6/0/160216AA-8445-480B-B60F-5C8EC8067FCA/WindowsAzure-SecurityPrivacyCompliance.pdf. Accessed 02 Sept 2018
28. Möller, B., Duong, T., Kotowicz, K.: This POODLE bites: exploiting the SSL 3.0 fallback. Security Advisory, September 2014. Accessed 02 Sept 2018
29. Neuman, D.C., Hartman, S., Raeburn, K., Yu, T.: The kerberos network authentication service (V5). RFC 4120, July 2005. https://doi.org/10.17487/RFC4120. https://rfc-editor/rfc/rfc4120.txt
30. Rescorla, E.: The transport layer security (TLS) protocol version 1.3. RFC 8446, August 2018. https://doi.org/10.17487/RFC8446. https://rfc-editor.org/rfc/rfc8446.txt
31. Rescorla, E., Dierks, T.: The transport layer security (TLS) protocol version 1.2. RFC 5246, August 2008. https://doi.org/10.17487/RFC5246. https://rfc-editor.org/rfc/rfc5246.txt

A Survey on Machine Learning Applications for Software Defined Network Security

Juliana Arevalo Herrera[1]([⊠]) [iD] and Jorge E. Camargo[2]([⊠]) [iD]

[1] Universidad Santo Tomás, Bogotá, Colombia
julianaarevalo@usantotomas.edu.co
[2] Universidad Nacional de Colombia, Bogotá, Colombia
jecamargom@unal.edu.co
http://www.unsecurelab.org

Abstract. The number of machine learning (ML) applications on networking security has increased recently thanks to the availability of processing and storage capabilities. Combined with new technologies such as Software Defined Networking (SDN) and Network Function Virtualization (NFV), it becomes an even more interesting topic for the research community. In this survey, we present studies that employ ML techniques in SDN environments for security applications. The surveyed papers are classified into ML techniques (used to identify general anomalies or specific attacks) and IDS frameworks for SDN. The latter category is relevant since reviewed paers include the implementation of data collection and mitigation techniques, besides just defining a ML model, as the first category. We also identify the standard datasets, testbeds, and additional tools for researchers.

Keywords: Software defined networks · Machine learning · Network security

1 Introduction

Separation of control and data planes is not a new idea, but only recently it has obtained high interest from the scientific community and commercial vendors with the popularization of Software Defined Networks (SDN). There have been several contributions to the technology, but it is still under development by the industry and academic community. In combination with other technologies such as Network Function Virtualization (NFV), SDN approach presents a solution to everyday problems existing in traditional networks like scalability and manageability issues. Additionally, it offers alternatives to monitor and control the traffic in the network, providing new possibilities for security applications. However, the de-facto protocol for control-data communication, OpenFlow [42], has been identified as a vulnerable solution [26]. It also presents additional security issues, as we will show in Sect. 3.

© Springer Nature Switzerland AG 2019
J. Zhou et al. (Eds.): ACNS 2019 Workshops, LNCS 11605, pp. 70–93, 2019.
https://doi.org/10.1007/978-3-030-29729-9_4

SDN definition is comprised of three layers. However, as technology develops, additional elements are required. In [18] Clark et al. propose a Knowledge Plane or KP as an individual entity for the network that aims to maintain a high-level view of the network and help in the operation, management, and improvement. Knowledge Defined Networking (KDN) [44], adds a knowledge plane (KP) to the SDN architecture, intending to integrate behavioral models and reasoning processes oriented to decision making. One of the tools to leverage the KP is Machine Learning.

Machine Learning is a powerful tool to provide cognitive capabilities for identifying security breaches. It has a significant improvement due to the processing and storage capabilities as well as the availability of large datasets. However, SDN is not broadly used in operative networks, though there is an important reference: Google's B4 [43] is a deployment of SDN over WAN network to connect several data centers. It included a switch design to handle the interconnection with traditional networks and ONIX [30] as the controller. It was proven to be a useful technique for the gradual integration of traditional to SDN infrastructure. The implementation did not present any contribution related to security, except for the use of the Paxos algorithm [33] for fault tolerance. Considering there are no available data on security research in SDN, obtaining realistic datasets for IDS becomes a challenge.

On [50], authors present an overview of the challenges and opportunities to use ML in new technologies such as SDN, however it is not exhaustive in the description or study. Other works such as [19, 32, 49, 61] have shown different ML techniques applicable to SDN anomaly detection but focus on the methods and lack of an analysis from the network security perspective.

In this paper, we present the most recent research (to the best of our knowledge) for network security in an SDN environment using ML techniques. Our motivation is to contribute to the creation of a KP for SDN, focused on security. The study presents the surveyed papers organized per network attacks, in contrast to other surveys related to ML methods used in SDN. It also shows the testbeds and datasets commonly used in the literature.

The rest of the paper is organized as follows. Section 2 presents the methodology used to select and classify the studies. Section 3 presents an overview of the SDN architecture and security issues. Section 4 presents the studies for ML-based techniques for IDS, only with a proposal of the detection model. Section 5 presents studies that include methods to collect data to feed the ML model, as well as mitigation schemes once the anomaly is detected. Section 6 aims to provide additional tools for researches with studies related to security, as well as used datasets and testbeds in the surveyed works. Finally, Sect. 7 concludes the presented survey.

2 Methodology

This survey focuses on the works that use machine learning (ML) including deep learning (DP) techniques to address security issues for software defined networks (SDN). We initially set the period of the publication to be used in the study as five years; however, during the first search within databases, we found valuable literature since the year 2013. Because of this, the publication period covers papers from that year until the beginning of 2019.

To search for papers for our study, we reviewed the IEEE Xplore, ScienceDirect, and Wiley databases, as well as Google Scholar to expand the scope to other repositories. The key-words used to conduct the study were: "SDN," "Security," "Machine learning," and "Deep learning." We combined the terms to create different search streams such as: ("SDN security" AND "machine learning"), ("SDN security" AND "deep learning"), OR ("SDN Security"). Only the titles of the studies were considered to select an initial list of 200 papers. Later, we classified the articles into those to be used in the survey and those to be excluded by reviewing the abstract, introduction and conclusion, only.

We selected papers that included all areas of the keywords (SDN, ML/DP, and network security), but also those that presented traffic classification or monitoring, since those methods are useful for securing the network. Out of the selected papers, we classified them into the following categories:

- Surveys
- Proposal for framework or security application
- Experiment of existing tools

Using this classification, we selected a total of 70 papers and excluded out of the initial list. These papers were reviewed in detail, and using them we identified other studies to be included.

3   SDN Architecture and Security

SDN was born out of the need to break the vertical integration of the network equipment. Its premise is to separate the control from the data plane, and the interface between them is OpenFlow (OF) [42] protocol, proposed in 2008, which leveraged its development. It also allows defining network functions (e.g., routing, firewall, load balancing, bandwidth optimization) as software applications that can run on top of the control plane. The architecture has three parts: data plane (composed of switches), the control plane (composed of one or more controllers), and application plane (composed of one or more network applications). Figure 1 shows an SDN architecture.

Within SDN, a flow is a set of packets with similar features that go from one endpoint (or group) to another endpoint (or group) in a single direction. Each flow has its entry in the flow table, which is a database within the switches consulted to determine what to do with each packet that arrives at the switch. The flow-tables are created by order of the controller. At the beginning of a transmission (new flow), the switch will receive a packet without an entry on the flow table. The OF protocol sends the "Packet_in" message, from the switch to the controller for analysis and definition of a new flow-table entry. The "Packet_in" is a particular feature that could become a vulnerability to the system. OF also defines the information collection, using a request from the controller that the switch answers with parts of the flow-table along with packet counters.

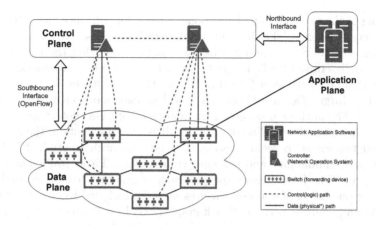

Fig. 1. SDN Architecture: Data plane, Control plane and Applications plane as its main components

This new paradigm represents a solution to several problems of traditional networks, such as manageability, configuration, scalability, and security. Under this perspective, a clear advantage for security with SDN is the ability to gather traffic information without additional elements. This is due to the centralized role of the controller, which communicates with the switches in the data plane. Proposals such as [8, 27, 47] take advantage of this ability to implement security functions such as Intrusion Detection Systems (IDS) and protection against Distributed Denial of Service (DDoS) within the network. SDN is, however, a model under development with open research lines and security challenges common to traditional networks, as well as unique to it. Different studies [7, 31, 57] presented analysis to network security from different viewpoints. A common conclusion is that security applications in SDN are still not mature enough for widespread implementation. A non-exhaustive review of SDN security architecture issues is presented below.

Kreutz et al. [31], created one of the firsts attempts to determine the vulnerabilities in SDN architecture. In this survey, the authors presented seven threat vectors: (1) Forged or faked traffic flow; (2) Attacks on vulnerabilities in switches; (3) Attacks on control plane communications; (4) Attacks on and vulnerabilities in controllers; (5) Lack of mechanisms to ensure trust between the controller and management applications; (6) Attacks on and vulnerabilities in administrative stations; and (7) Lack of trusted resources for forensics and remediation. Other studies [56, 71] also used this scheme to analyze SDN security. The paper also proposes the mechanisms required to secure a controller: Replication, Diversity, Self-healing mechanisms, Dynamic device association, Trust between devices and controllers, as well as between controllers and applications, Security domains, Secure components, and Dependable maintenance of software.

The first attack vector was exploited in [58]. Initially, they detect if a given network uses SDN by comparing the response times. If it is SDN, at the beginning of the transmission the response time is longer, since the network has a "flow setup" latency. The times have subtle differences, so the authors present a solution with an SDN scanner. After the confirmation that the network is an SDN, specially crafted traffic is sent to the network to cause data plane resource consumption or Distributed Denial of Service (DDoS) attacks.

Moving Target Defense (MTD) is a widespread approach used by several studies. In [17], authors proposed a framework to prevent, detect and mitigate attacks. The research was directed to virtualized environments in the cloud and presented two areas to secure resources. First, the authors studied MTD for network programmability and software vulnerability. Then, traffic engineering was reviewed. The latter allows the provision of different tenants securely. For the former one, a set of countermeasures must be included to enforce after the detection and analysis with an attack graph (AG) based vulnerability analysis.

The same approach was studied in detail in [16]. The authors presented AG techniques to reconfigure the network automatically and used MTD as a countermeasure. However, it does not present information on the attack detection but assumes the intrusion detection already in place. It still needs a phase for attack analysis in which ML could be used.

Few studies present machine learning solutions for the SDN architecture security problems identified by [31]. However, some works suggest the possibility to use it. In [64], authors presented three levels of complexity to use cognition: Reactive reasoning (rule-based reaction), Tactical reasoning (Profiling based on classification with dynamic multi-objective optimization), and Strategical reasoning (Anticipation with online multi-objective optimization). The study proposes to formulate optimization functions related to the security concerns in the network.

On the other hand, in [29] authors presented a framework to provide autonomous response and mitigation against attacks in an SDN/NFV network. The approach is called SARNET and has a transverse loop with five stages: Detect, Analyze, Decide, Respond, and Learn. An essential contribution of the study is the definition of an efficiency estimation that allows measuring the performance of the proposed framework. A group of simulations of different attacks (UDP DDoS attack, CPU utilization attack, Password attack) showed that the efficiency measure helps in selecting the best countermeasure. Within all the loop, it is suitable to use ML, and the authors present it as future research.

As presented, there is not extensive research to secure the SDN architecture using ML. However, SDN architecture can leverage network security since it allows the managers to know, rather than infer, the specific status of the network. OF gives the opportunity to collect statistics and traffic information that could be used to identify anomalies, intruders or configuration failures within the controller, devices or applications.

These abilities present the possible implementation of security applications on top of the SDN architecture. They are also leveraged by the use of Network Function Virtualization (NFV). NFV intends to apply IT virtualization technology for networking functions [15], and the objective is to break the dependence of hardware. In this scenario, security applications can be implemented on commodity devices, and the necessity of specific equipment could be eliminated.

In the following sections, we will discuss the different proposals to use SDN as a mean to improve network security. Our approach is to analyze the use of machine learning to achieve the desired result. As presented in Fig. 2, we classified the papers into Type 1: ML-based intrusion detection Systems in SDN, and Type 2: ML-based intrusion detection Systems in SDN. In the first case, the sub-classification depends on the type of detected attack. In the second case, the sub-classification depends on the data collection method to feed the ML-Model.

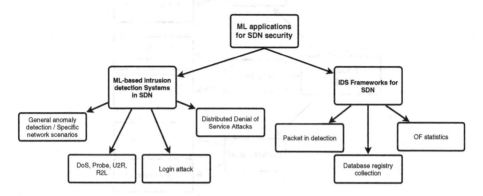

Fig. 2. Classification of studies in the survey

4   ML-Based Intrusion Detection Systems in SDN

Intrusion detection systems (IDS) are one of the most widespread applications for security in SDN. Since OF provides traffic statistics using the messages "StatsRequest" and "StatsResponse," it becomes a compelling tool to identify anomalies and intruders.

Fundamentals of IDS operations apply equally for traditional and SDN environments. Considering the location of the method IDS techniques can be divided into Network IDS and Host IDS. The former performs intrusion detection by analyzing the overall situation of the network. On the other hand, HIDS is host-based detection that monitors the operation of a particular device.

As detection mechanisms, IDS employ two types of strategies: (1) Traditional, signature-based detection that compares data to an existing database; and (2) Anomaly-based detection, which identifies odd-behaviour traffic, and can make use of ML techniques for better results. Examples of IDS proposals with the traditional approach in SDN are [14,40,72]. For instance in [40] of the first attempts to identify anomalies issues leveraging on SDN. The intention was to determine the main security issues related to the cloud computing environment to propose an SDN-based approach that allows the network to react in case of an attack. On the other hand, in [72] the authors proposed a Deep Packet

Inspection system for network intrusion detection and prevention using NFV. It was implemented, and it presented reasonable performance. Finally, authors on [14] proposed to detect and mitigate anomalies in SDN, with a statistical approach for detection. A definition of a "normal traffic" profile is the base for the statistical analysis.

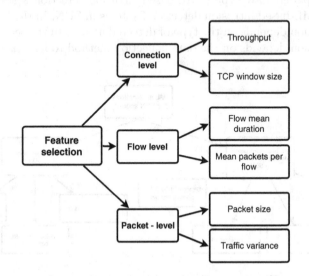

Fig. 3. Features to select in network traffic

At the packet level, the information can be statistical for the network and related to packet size, variance, root mean square. It is useful to characterize traffic in the network, for example with the Hurst parameter H, used to measure the self-similarity and burstiness (the burstier the traffic, the higher H) [37]. Flow and connection level features are most commonly used in SDN, as we will show below. Examples of each level are presented in Fig. 3.

In the sections below, we will present the surveyed papers and a summary in Table 1.

4.1 General Anomaly Detection

Some studies propose general anomaly detection with ML. For example, in [21,22], authors present IDS with deep learning techniques applied to SDN environments. Both studies implemented the IDS as a component of the control plane, instead of deployment as an application. The location allows interacting directly through the network hence protect the controller itself. In [21], they presented a general SDN environment with unsupervised learning. The approach is to use an autoencoder, which has two phases (encoder + decoder) to detect and minimize the reconstruction error for each test sample. The development library was Tensorflow although it is not clear what was the used dataset. The second study presents a secure framework for IoT based on SDN with a brief review of the security in SDN architecture, but also presents a ML-based IDS. It uses deep learning with a Restricted Boltzmann Machine (RBM). For simulation, the authors focused on the detection model with Tensorflow, and the dataset used was KDD99. The proposed algorithm showed 94% of accuracy.

Authors in [60] present a proposal for both IDS and an action triggered by it: Moving Target Defense. They created a simulated network to obtain data for the training (about 40,000 packets). For the architecture, they presented a neuro-evolutionary model as a light-weight detector that allows real-time operation. To achieve it, they developed two distinctive detectors, one per each type of attack to identify: DDoS and worm. To combine the detectors, authors use Neuro-evolution of Augmenting Topologies (NEAT), an approach to neuro-evolution with crossover context.

4.2   Specific Network Scenarios

There are also proposals for specific network scenarios. That is the case of [73] that presents the implementation of ML-based IDS in optical SDN, and [55] that proposes an scenario of Intelligent Transport Networks. The study in [73] starts by surveying the attacks in control plane and categorize them into unauthorized access, data leakage, data modification, denial of service, and security policy misuse. Since the scenario is optical networks, anomaly detection must consider features related to optical links. Examples are average bandwidth usage, frequent source and destination nodes, average route length, and modulation formats. The possible attacks in this type of network include light-path creation, modification, and deletion (all directed to the link-data layer of the OSI model in optical networks). The first detection methods are point-anomaly-based, as a data instance represented by a point is outside a common region of normal behavior. It uses an algorithm created by the authors to calculate a probability. The second is a sequence-anomaly based method where anomalies occur together as a sequence and use an improved cumulative sum approach. For testing, the authors use NSFNET topology with an owned dataset, and the results present an average detection accuracy of 85%.

On the other hand, [55] presents the cross-fire attack in ITS. The attack consists of a large number of compromised nodes that generate coordinated and low-intensity traffic to disconnect victims (hosts or links) from the network. A ML approach is used to classify the coordinated attacks using three deep learning algorithms: (1) Artificial Neural Networks (ANN); (2) Convolutional Neural Networks (CNN); and (3) Long Short-Term Memory (LSTM) networks. The authors created a testbed in mininet [65] to generate a dataset of their own, with increased traffic for the compromised nodes. They later used this dataset to train and test the model. The results proved the efficiency of the proposal with a slight reduction of performance when the speed of the vehicles increases. A highlight from the study is that it presents the training time and it is about 100 seconds for each algorithm. The short time allows the system to be re-trained as necessary.

4.3 Login Attack

From the surveyed papers, we only found one that addresses login attacks in [46]. The proposal includes defining security rules on the SDN controller to identify and block that type of threat. The study presents the feasibility with the use of four ML techniques: C4.5, BayesNet (BN), Decision Table (DT), and Naive-Bayes (NB). The intention is to give the network the ability to act against a chain of attacks from multiple IP addresses used by each attacker. The used features for the models are attacker IP, attacked host, number of attempts in an attack, and timestamp. The study shows that even a small probability of attack should not be ignored and security rules on the SDN controller must be accordingly modified. For experimentation, the "long tail" dataset was used [23].

4.4 DoS, Probe, U2R and R2L

The studies presented in this section address four kinds of attacks: DoS, Probe, User to Root (U2R), and Remote to local (R2L). The common characteristic between all of them is the dataset used: NSL-KDD [13] that classifies the attacks in the aforementioned categories.

In [62], authors proposed the use of deep neural networks to detect anomalies based on six flow-based features regarded as suitable for SDN: duration, protocol_type, src_bytes, dst_bytes, count and srv_count. The authors trained and tested the model, and compared their proposal with other algorithms such as J48, Naive Bayes (NB), NB Tree, Random Forest (RF), Random Tree (RT), Multi-layer Perceptron, and Support Vector Machines (SVM). The paper states the potential of deep learning for the flow-based anomaly detection system. Authors also argue that ML is not fully developed.

In [35], an study of nine ML classifiers with supervised machine learning approaches is presented. They perform tests for accuracy, false alarm rate, precision, recall, f1-measure, the area under the curve (ROC), execution time and Mc Nemar's test. The tests were made with Principal Component Analysis (PCA) for dimensionality reduction with NN, Linear Discriminant Analysis (LDA), DT, RF, Linear SVM, K Nearest-Neighbour (KNN), NB, Extreme Learning Machine (ELM), AdaBoost, RUSBoost, LogitBoost, and BaggingTrees. The results showed that DT, bagging and boosting approaches had better performance than the rest. The selected features were a subset of the features of the dataset, excluding content features.

The same authors proposed in [34] a 5-level hybrid classification system for IDS inspired in the work presented in [9], in a not-SDN network. The paper aims to use flow-statistics provided by the controller to develop a NIDS. The classification methods used are the kNN in the first level, ELM for the second level, and Hierarchical Extreme Learning Machine (H-ELM) for the rest. Each level detects a type of attack using the same features selected in [62]. The system was implemented as a module of POX controller instead of a function of the application plane, for scalability purposes. The approach for selecting these features is the easiness to get them directly from the controller. The results presented showed improved accuracy, compared to other techniques.

Authors in [53] also place their IDS in the control plane. The technique is a meta-heuristic Bayesian network to classify traffic, and the dataset is NSL-KDD. The proposed process includes a phase of feature selection and extraction to optimize the classifier that consists of the fitness evaluation of the extracted features. It is later fed to the Bayesian classifier. The proposed algorithm is compared with seven other approaches and showed the best overall efficiency for the performance measures with a total of 82.99%.

4.5   Distributed Denial of Service Attacks

Although several of the previous studies consider DoS attacks, they are presented as part of a greater range. In this section, we present studies that investigate specifically DDoS attacks for two reasons. The first one is that a large section of studies for IDS focuses on DDoS attacks. Secondly, with the perspective of the Internet of Things (IoT) and recent threats such as Mirai botnet [28] it is worth to consider the attack individually.

In [68], authors present a specific application for anomaly detection using SDN as a solution to solve scalability challenges. The scenario is a Wireless SDN enable E-Health system. The main feature of this type of network is the massive machine-type communications (mMTC) in which human interaction is minimal. The ML technique used is contrastive pessimistic likelihood estimation (CPLE) for semi-supervised operation with offline training. The intention is to perform online testing to allow running localized detection within the devices to avoid the need to frequently collect network traffics at the controller to update the anomaly detection model. The features used for the classifications are the same defined by [62].

In [11], authors provided an overview of the use of ML for IDS in SDN. The study investigates five ML techniques to mitigate intrusion and DDoS attacks (Neural networks, support vector machine, genetic algorithms, fuzzy logic, Bayesian networks, and decision tree). The authors theoretically analyzed each method and generated a comparison scheme that presents the pros and cons of the techniques. The paper serves as an initial review to select the best approach, according to the needs of the system. However, it does not proposes or test any model.

An analysis of SVM and comparison with other techniques for DDoS detection in SDN is presented in [27]. The paper briefly discussed the types of DDoS attacks and security threats to the controller in SDN. Later, the paper gave four SMV methods and the system description. The 1999 and 1998 DARPA datasets were used for training and testing (about 50/50 ratio), and the technique was compared with RBF, Naive Bayes, Bagging, J48, and Random Forest methods. Accuracy was highest for the proposed SMV with 95%.

In [70], the authors proposed a learning algorithm based on Support Vector Classifier (SVC), leveraged on an Iterative Dichotomiser 3 (ID3) decision tree for feature selection. The model was evaluated in a software testbed with three main components (1) Open vSwitch as a virtual switch, (2) Ryu as the controller, and (3) sFlow Toolkit for data collection. The used dataset is KDD-Cup 1999.

A Dirichlet Process Mixture Model is used in [6], to mitigate DNS-based DDoS attack. Authors used an owned dataset created from the technique to generate them presented in [59].

In [52], authors present an IDS system to identify DDoS attacks. They compare three methods: Naive Bayes, K-Nearest Neighbor (KNN BEST), and Support vector machine (SVM) with an accuracy of 97%, 83%, and 83%, respectively. The features considered as inputs are the number of Packets, Protocol, Delay, Bandwidth, Source IP, and Destination IP. For testing, they use an owned dataset.

In [24], authors present a proposal to improve resiliency in an SDN network, by detecting DoS attacks, specifically SYN flood attack. For classification, the study shows three different techniques: DT, SVM, and NB. The results presented over 99% accuracy, recall, and precision for DT. Dataset KDD 99 is used in the study with the features source IP address, destination IP address, source port, destination port, and protocol. They are later reduced using PCA.

Authors in [48] present an approach to detect and classify DDoS attacks in a cloud environment. For it, they use a two-stage ensemble learning scheme with multivariate Gaussian and Bayesian techniques. The employed features are src_ip, dst_ip, no_of_packets, spoof_dst_ip, blacklist_ip. Although the study is composed of complementary elements to the ML technique, it does not directly try to secure and SDN. Instead, it defines the steps to protect the cloud infrastructure (Virtual machines, orchestrators, etc.).

The previous works were the application of ML techniques for IDS. However, they do not consider implementation issues within the network. In Sect. 5, we present a set of works classified as "frameworks," since they include considerations such as collection and mitigation methods.

4.6 Techniques Comparison

Considering the broad spectrum of cyber-security attacks is noteworthy to have just six specific attacks (DoS, DDoS, Probe, U2R, R2L and login). Even though SDN is an innovative paradigm, we could expect every type of known attack used against an SDN. Also, the research community should prepare to deal with new adapted attacks. It is essential to review how to adapt current techniques to detect, mitigate and prevent different attacks in SDN. Several of the attacks already are recognized using ML techniques applied to them in traditional networks.

Table 1 shows that ML techniques used are very diverse. Most of the papers (9 out of 16) use a single ML technique. The others use at least two methods with one of two approaches: comparison between techniques or combination of them to improve the anomaly detection. Artificial Neural Networks were used in 50%

Table 1. ML techniques proposals for anomaly detection in SDN

Ref.	Detected attack	Detection method	Feature selection	Training dataset
[22]	General anomaly	RBM	41 Features	KDD-Cup 1999
[21]	General anomaly	Autoencoder	41 Features	KDD-Cup 1999
[73]	Optical network	Point anomaly: probability-based. Sequence anomaly: CUSUM	Related to optical links. (e.g. bandwidth, source and destination nodes, route length, and modulation formats)	NSFNET
[60]	DDoS and worm	NEAT	3 packet-level features	Owned: 800000+ packets
[35]	DoS, Probe, U2R, R2L	DT, ELM, NB, LDA, NN, SVM, RT, KNN, AdaBoost, RUSBoost, LogitBoost and BaggingTrees	Subset of features and Principal Components Analysis (PCA) approach	NSL-KDD
[53]	DoS, Probe, U2R, R2L	MHBNC	Preprocessing + feature extraction	NSL-KDD
[62]	DoS, Probe, U2R, R2L	DNN	6-flow-based features	NLS KDD
[34]	DoS, Probe, U2R, R2L	kNN, ELM, and H-ELM for the rest	6-flow-based features	NSL-KDD
[68]	DoS, Probe, R2L and U2R	CPLE	6 features vs 41 features	NSL-KDD
[46]	Login	C4.5, BayesNet, Decision Table (DT), and NB	4-attack-based features	LongTail.
[55]	Crossfire	ANN,CNN,LSTM	3-flow-based features	Owned
[70]	SYN Flood DDoS	SVC	ID3	KDD-Cup 1999
[27]	DDoS	SVM	Grid search method	1999& 1998 DARPA
[52]	DDoS	NB, KNN BEST and SVM	6 fixed features. 6000 data samples	Owned
[48]	DDoS	Ensemble learning with multivariate Gaussian and bayesian	5 flow-based features	Owned
[24]	DDoS	DT, SVM, and NB	4-flow-based features and reduce space withPCA	KDD-Cup 1999

(RBM, NEAT, Generic NN, KNN, ANN, CNN). Another common approach in the reviewed papers is the use of Support Vector Machines. Several articles also presented a Naive Bayes method. However, it was only part of a comparison to other techniques.

Finally, considering feature selection, we found it very diverse. However, in [62] the authors presented a set of six features that were used in four studies, regarded as suitable for SDN. On the other cases, the technique or definition of the features to be included in the ML model was independently selected.

5   IDS Frameworks for SDN

The implementation of the ML techniques for IDS needs to consider articulation with the network environment. That is, define how to collect the data for analysis, as well as mechanisms to activate in case of anomaly detection. For collection, we found three main sources of data to feed the ML model: (1) Statistic collection with OF methods [8,36,38], (2) Getting a copy of the flow table from the switches [45], and (3) With packet-in messages [20,63,67].

Regarding the mitigation, the typical method is to define a module at the control plane (next to the controller) or a dedicated application in the application plane that affects the OF tables of the switches.

In the paragraphs below, we will present the frameworks found in the survey and their main considerations, in contrast to the previous section (studies of the single ML model). The studies are organized regarding the collection method.

5.1   Frameworks Description

Authors in [8] present a system that applies Machine Learning (ML) classification algorithms to detect DDoS attacks. They also propose two defense mechanism for specific SDN attacks: miss-behavior attack and new-flow attack. The first refers to the attack directed to a target using an existing, validated flow. The second exploits the packet-in vulnerability to create a DoS attack. Both are statistical-analysis based. Regarding the DDoS detection mechanism, the system uses a ranker algorithm, a genetic algorithm, and a greedy algorithm for feature selection and Sequential Minimal Optimization (SMO) for classification. The achieved accuracy is 99.40%.

OF statistics are also used in [38] with a 5G scenario implemented with SDN. The study presents Random Forest classifier for feature selection and combines k-means++ with Adaboost for flow classification. The former creates two clusters, which most probably represent the normal and abnormal instances and the later further partition the anomaly clusters into four main classes of attacks. The techniques are part of a complete architecture for ML-based IDS within the SDN scheme. It includes modules in each plane of SDN to allow the collection of data and mitigation action. The ML techniques used are varied and does not evaluate the classification algorithms, but the combination of them with the feature selection techniques. The combinations in the study are RF-KA, RF-GDBT, RF-DT, RF-SVM, Tree-KA, Fisher-KA, and ReliefF-KA. The study presents an analysis of these combinations in an environment that balance the attacks (over-sample the minority intrusion such as R2L, and under-sample the majority intrusions such as DDoS). For evaluation, the study uses KDD-Cup 1999 [66]. Two relevant conclusions from the study are: (1) Feature selection is critical for better accuracy and lower false rate; and (2) The sampling technique could improve the detection accuracy of minority intrusions dramatically while maintains a reasonable detection rate of the majority ones.

In [36], authors present a framework to use ML for IDS. They propose a NIDS over SDN architecture in which the packets from the switches are captured on a computer with many network cards that act as OpenFlow vSwitch. It sends the Ethernet packet to a Feature Extractor module that analyzes them and extracts 25 features, depending on the transport protocol (TCP, UDP, ICMP). Later the C4.5 algorithm classifies packets for malicious activity. For testing, the authors used the 1999 Darpa dataset [39], and they showed detection of DoS and Probe attacks at high precision. They also proposed and tested a network topology to generate real traffic.

The second type of collection method is to obtain flows from the data plane, using the forwarding.l2_learning Method provided by POX. The technique is used by [45] in combination with an unsupervised RBM algorithm with 92% accuracy. The training method is based on Contrastive Divergence (CD), and the features used for the model are flow-level, and connection-level: total number of packets transmitted (ToP), the ratio of source and destination bytes (RoSD), and connection duration time (CT).

Another technique to collect data is the use of packet-in messages of OF. The method is proposed as part of the framework DaMask in [67]. Even though it is presented as DDoS detection, the study does not present the ML detection technique. According to that, the architecture could be implemented in other types of attacks. The primary goal is to apply DaMask to a cloud computing environment from an enterprise view, which is inherently different than a network. The identified differences are: (1) Control of the computational resources are out of hands of the defender (provider's responsibility); (2) Fast and straightforward resource allocation generates constant topology changes to adapt to; and (3) Network resources are shared with all other users of the cloud, which requires separation mechanisms not considered in traditional DDoS. To answer the requirements, the authors created a three-layer framework (one per each plane in SDN). The system has two main modules (attack detection and attack mitigation) at the application level. For feature selection, they used the Chow–Liu algorithm, and the attack detection is made with a graphical model. The testing was done with the UNB ISCX [59] dataset. As a result of the evaluation, the authors concluded the proposed framework requires little effort from the provider for implementation.

Packet-in detection as a collection method is also used in [20], in combination with a neural network for detection of DDoS attacks. The solution consists of four mechanisms: attack detection trigger, attack detection, attack traceback, and attack mitigation. The study of the detection trigger (when to start the detection process) and traceback (find the source of the attack) are differentiators for this proposal. Similarly to other proposals, [63,67] the authors selected an abnormal detection of packet-in messages as a trigger to start the detection mechanism (Backpropagation neural network BPNN). It has one input layer (five neurons), one hidden layer (ten neurons) and one output layer (one neuron). On the other hand, the backtracking mechanism seeks for the path followed by the malicious flow by marking the switches, which allows identifying the source. The mitigation method creates new flow entries with the highest priority to drop the traffic directed to the target, and use OpenFlow modification message to clean the flow tables. The study presents the results based on the performance of the detection trigger but not the BPNN classification.

Finally, authors in [63] also use packet-in detection as a collection method and present a Gated Recurrent Unit Recurrent Neural Network as part of a framework for IDS. The detector is implemented as part of the control plane, next to the controller. For this case, the feature srv_count is changed for the dst_host_same_src_port_rate, although they used the same features and dataset of their previous work [62]. The proposal presented low processing impact on the controller and a detection rate of 89%.

5.2 Frameworks Comparison

In Table 2 we present the surveyed frameworks. Only seven (7) out of the studied papers, presented a complete framework to implement in a network. The elements identified in these papers to classify them as frameworks are the description of collection and mitigation methods. They are applied before and after the detection mechanism and provide a clear architecture to deploy the solution in a functioning network.

We identify three types of collection methods: OF statistics, database copy with forwarding.l2-learning command, and packet-in. All of the methods are based on OF possibilities. However, there is diversity in SDN implementation, and it is essential to define other alternatives for other scenarios. An appealing option is sFlow [4], a monitoring tool for packet sampling with an analysis module.

For mitigation, papers [36,45] do not provide a proposal. Frameworks [8,20, 38,63] base their technique on the use of OF, with table modification on the data plane. The proposals consider an additional module in the controller to handle the changes.

However, the proposal in [67], DaMask, presents an architecture in which the mitigation is located on the application layer of SDN. That approach would allow some flexibility for the deployment of the design.

Table 2. Framework proposals to use ML in anomaly detection in SDN

Ref.	Collection method	Mitigation method	Detection method	Feature\selection	Metric	Training dataset	Scenario
[8]	OF based, packet recieved counter	Openflow table change from app level	Sequential Minimal Optimization (SMO)	Ranker, genetic, and greedy algorithms	Accuracy: 99.40%	NLS-KDD	SDN
[38]	OF based, regular intervals	Dedicated module to give instructions to switches	(1) Statistical (2) ML techniques (GBDT, DT, SMV, KA)	RF, Tree, Fisher, ReliefF	Accuracy: Probe 97.96% DoS 99.97% U2R 68%, R2L 65.5%	KDD-Cup 1999	5g networks
[36]	OF based, not detailed	Not defined	C4.5	14 derivated from transport features from a basic set of 9	IPS alert: 60%	1999 Darpa	SDN
[45]	Flows from the data plane switches saved in POX database (forwarding.l2-learning)	Not defined	Restricted Boltzmann Machine based	Restricted Boltzmann Machine	accuracy was 92%	Owned. Undeclared features	SDN
[63]	Packet-in detection	Openflow table change	Gated Recurrent Unit Recurrent Neural Network (GRU-RNN)	Fixed (six features)	Accuraccy 89%	NSL-KDD	SDN
[20]	Packet-in detection (Abnormal messages trigger ML detection)	Openflow table change from app level	Backpropagation neural network BPNN	Not declared	Time, cpu use and traffic due to the trigger method for detection	Owned. Undeclared features	SDN
[67]	Packet-in detection	Managed from app layer	Graph method (not specified ML)	Chow–Liu algorithm	Detection rate (%) Basic 74.02, Local 86.56, Global 89.30	UNB ISCX	Cloud

6   Complementary Proposals, Datasets and Testbeds

To identify open research problems, as well as the primary tools, we present in this section other ML studies related to security, datasets used from the surveyed studies and used testbeds in the cases a network simulation or emulation was created, that is only for the frameworks.

6.1   Other ML Studies Related to Security

Additional to the use of ML for IDS, we identify other studies to consider. On the first place, we recognize the issue related to adversarial machine learning, which was addressed by Nguyen in [49]. The author presented a cyber kill-chain directed to attack machine learning models. The study provides an analysis of the current use of ML in SDN security as well as attacks directed to ML models such as equation-solving, model inversion, pathfinding, and others. It then presents the cyber kill chain, composed of seven steps: (1) Recon; (2) Weaponization; (3) Delivery; (4) Exploitation; (5) Installation; (6) Command and control; and (7) Action. The paper concludes with four recommendations to use ML in network security: (1) Invest time and effort in the threat models while designing ML solutions; (2) Make the ML model auditable; (3) Follow a secure development process; and (4) Produce an initial operational cost model.

An open, available implementation of ML techniques for IDS is [1]. Authors in [41] perform tests on the platform and concluded that the ML algorithm a large training dataset to reduce the false positives. They also present the possibility to create poisoning attacks to cause miss-classifications.

Additionally, it is important to identify tools that could be used in the analysis of traffic. Studies such as [10,12,25,54,69] present ML-based traffic classifiers to identify applications or flow features in different SDN scenarios. Although the proposals are not specific for security, they might leverage the implementation of security applications.

6.2   Datasets and Testbeds

Regarding datasets, from Tables 2 and 1, we identify a total of six public datasets used on the studies. In Table 3 we present the available datasets (items 1 to 6) and also a type that was created by the researches (item 7). The last two columns of the table indicate how many studies use a particular dataset for Type 1 studies (Sects. 4.1 to 4.5), and the second represents Type 2 studies (frameworks presented in Sect. 5).

It is noteworthy that most of the studies use similar datasets, which could cause the same bias issues in the models. Twelve studies use the KDD-Cup 1999 and NSL-KDD datasets that are 20 and 10 years old respectively. Even though they are used extensively in the research community, it is crucial to consider that attacks become more and more sophisticated every day. Besides the owned datasets, LongTail is the newest, but a single study uses it.

Table 3. Datasets used for ML-based IDS in SDN

Item	Dataset	Year	Studies	
			Type 1	Type 2
1	DARPA 99	1999	1	1
2	KDD-Cup 1999	1999	4	1
3	LongTail	2015	1	0
4	NSFNET topology	NA	1	0
5	NSL-KDD	2009	5	2
6	UNB ISCX	2012	0	1
7	Owned	NA	4	2

Regarding the datasets generated by the authors (classified as owned), standard tools are Mininet, Scapy [5], Distributed Internet TrafficGenerator (D-ITG) and the DDoS attack tool TFN2K.

A common approach for creating datasets is to use the guide provided in [59]. The study presents a systematic approach to develop datasets although it is not focused on SDN.

A more modern methodology is presented in [51]. The paper describes a controlled environment to experiment and create datasets for training supervised ML components and validate supervised and unsupervised solutions. The intention is to fill two gaps: (1) The need for threat data generation; and (2) Lack of new datasets to design, train and validate ML models, instead of the old, overused dataset. That is the case of the NSL-KDD. The proposal is an application of NFV/SDN than ML. It presents, however, the possibility to obtain data to be used in these type of systems.

Table 4. Testbeds used for ML-based IDS in SDN

Framework	Testbed
[8]	Emulation on mininet with pox controller and four OVS switches
[20]	Emulation on mininet with RYU controller and 25 switchES with 200 hosts (2 different computers)
[36]	Network implementation with Opendaylight controller and single computer with many network cards acting as an Openflow vSwitch.
[38]	Not defined
[45]	Emulation on mininet with POX controller and one switch with 5 hosts
[63]	Emulation with Cbench with POX controller
[67]	Emulation on mininet implemented in public cloud (AWS EC2) and extended in a privated cloud with Floodlight controller one switch and two hosts. One of the hosts is a web server

On the other hand, for testing of the complete frameworks, the most common tools was Mininet. Authors also used Open VSwitch [3], and Cbench [2] for emulation, as well as network implementation in the case of [36]. Authors in [36], used public cloud environment AWS EC2 in combination of an emulated private cloud.

We present the description of each testbed in Table 4.

7   Conclusion

We present the state of the art of ML-based SDN security proposals. The classification into ML techniques and frameworks allows identifying that very few attacks are being studied in this context. Considering the broad spectrum of cyber-security, there should be more work on different kind of attacks. Additionally, most of the proposals do not include collection techniques to feed the ML model, or mitigation methods to act after detection. There is a need to define specific schemes to implement the ML techniques in SDN. We also identify the need to use updated and SDN-specific datasets that allow creating models to fit actual networks and current attacks. Finally, we present the typical testbeds for the proposals that include network implementation, where there is no implementation on any operative network. This survey allows scholars to find out new research directions that address open problems in SDN security at different levels. There are also opportunities to involve ML techniques to solve such problems.

We also show in this paper that the use of ML techniques in SDN scenarios is an interesting topic for the research community. However, some aspects receive little attention and could be studied further. One key finding is related to the absence of enough open datasets that can be used to compare new methods. From the networking perspective, there is a lack of a comprehensive attack detection that considers a broad spectrum.

As future work, we want to extend the analysis of the ML techniques used in the reviewed papers with a more detailed study.

References

1. Apache spot. http://spot.incubator.apache.org
2. CTools:CBench - cTuning.org. http://ctuning.org/wiki/index.php/CTools:CBench
3. Open vSwitch. https://www.openvswitch.org/
4. sFlow.org - Making the Network Visible. https://sflow.org/
5. Welcome to Scapy's documentation!—Scapy 2.4.2-dev documentation. https://scapy.readthedocs.io/en/latest/
6. Ahmed, M.E., Kim, H., Park, M.: Mitigating DNS query-based DDoS attacks with machine learning on software-defined networking. In: Proceedings - IEEE Military Communications Conference MILCOM (2017). https://doi.org/10.1109/MILCOM.2017.8170802

7. Ali, S.T., Sivaraman, V., Radford, A., Jha, S.: A survey of securing networks using software defined networking. IEEE Trans. Reliab. **64**(3), 1086–1097 (2015). https://doi.org/10.1109/TR.2015.2421391

8. Alshamrani, A., Chowdhary, A., Pisharody, S., Lu, D., Huang, D.: A defense system for defeating DDoS attacks in SDN based Networks. In: Proceedings of the 15th ACM International Symposium on Mobility Management and Wireless Access - MobiWac 2017, pp. 83–92. ACM Press, New York (2017). https://doi.org/10.1145/3132062.3132074

9. Al-Yaseen, W.L., Othman, Z.A., Nazri, M.Z.A.: Multi-level hybrid support vector machine and extreme learning machine based on modified K-means for intrusion detection system. Expert Syst. Appl. **67**, 296–303 (2017). https://doi.org/10.1016/j.eswa.2016.09.041

10. Amaral, P., Dinis, J., Pinto, P., Bernardo, L., Tavares, J., Mamede, H.S.: Machine learning in software defined networks: data collection and traffic classification. In: 2016 IEEE 24th International Conference on Network Protocols (ICNP), pp. 1–5. IEEE, November 2016. https://doi.org/10.1109/ICNP.2016.7785327

11. Ashraf, J., Latif, S.: Handling intrusion and DDoS attacks in software defined networks using machine learning techniques. In: 2014 National Software Engineering Conference, pp. 55–60. IEEE, November 2014. https://doi.org/10.1109/NSEC.2014.6998241

12. Bakhshi, T.: Multi-feature enterprise traffic characterization in openflow-based software defined networks. In: 2017 International Conference on Frontiers of Information Technology (FIT), pp. 23–28. IEEE, December 2017. https://doi.org/10.1109/FIT.2017.00012. http://ieeexplore.ieee.org/document/8261006/

13. Canadian Institute for Cybersecurity: NSL-KDD Datasets. https://www.unb.ca/cic/datasets/nsl.html

14. Carvalo, L.F., Abrao, T., de Souza Mendes, L., Proença, M.L.: An ecosystem for anomaly detection and mitigation in software-defined networking. Expert Syst. Appl. **104**, 121–133 (2018). https://doi.org/10.1016/j.eswa.2018.03.027

15. Paper, N.W.: Network functions virtualisation: an introduction, benefits, enablers, challenges & call for action. Issue 1 (Technical report, ETSI) (2012)

16. Chowdhary, A., Pisharody, S., Huang, D.: SDN based Scalable MTD solution in cloud network. In: Proceedings of the 2016 ACM Workshop on Moving Target Defense - MTD 2016, pp. 27–36. ACM Press, New York (2016). https://doi.org/10.1145/2995272.2995274

17. Chung, C.J., Xing, T., Huang, D., Medhi, D., Trivedi, K.: SeReNe: on establishing secure and resilient networking services for an SDN-based multi-tenant datacenter environment. In: 2015 IEEE International Conference on Dependable Systems and Networks Workshops, pp. 4–11. IEEE, June 2015. https://doi.org/10.1109/DSN-W.2015.25. http://ieeexplore.ieee.org/document/7272544/

18. Clark, D.D., Partridge, C., Ramming, J.C., Wroclawski, J.T.: A knowledge plane for the internet. In: Proceedings of the 2003 Conference on Applications, Technologies, Architectures, and Protocols for Computer Communications - SIGCOMM 2003, p. 3. ACM Press, New York (2003). https://doi.org/10.1145/863955.863957

19. Coughlin, M.: A survey of SDN security research. Technical report. http://ngn.cs.colorado.edu/~coughlin/doc/a_survey_of_sdn_security_research.pdf

20. Cui, Y., et al.: SD-Anti-DDoS: fast and efficient DDoS defense in software-defined networks. J. Netw. Comput. Appl. **68**, 65–79 (2016). https://doi.org/10.1016/J.JNCA.2016.04.005. https://www-sciencedirect-com.ezproxy.unal.edu.co/science/article/pii/S1084804516300480

21. Dawoud, A., Shahristani, S., Raun, C.: A deep learning framework to enhance software defined networks security. In: 2018 32nd International Conference on Advanced Information Networking and Applications Workshops (WAINA), pp. 709–714. IEEE, May 2018. https://doi.org/10.1109/WAINA.2018.00172. https://ieeexplore.ieee.org/document/8418157/

22. Dawoud, A., Shahristani, S., Raun, C.: Deep learning and software-defined networks: towards secure IoT architecture. Internet Things **3–4**, 82–89 (2018). https://doi.org/10.1016/J.IOT.2018.09.003.   https://www.sciencedirect.com/science/article/pii/S2542660518300593

23. Eric Wedaa: LongTail (2015). http://longtail.it.marist.edu/honey/dashboard.shtml

24. Gangadhar, S., Sterbenz, J.P.G.: Machine learning aided traffic tolerance to improve resilience for software defined networks, pp. 1–7 (2017)

25. He, L., Xu, C., Luo, Y.: vTC. In: Proceedings of the 2016 ACM International Workshop on Security in Software Defined Networks and Network Function Virtualization - SDN-NFV Security 2016, pp. 53–56. ACM Press, New York (2016). https://doi.org/10.1145/2876019.2876029

26. Kloti, R., Kotronis, V., Smith, P.: Openflow: a security analysis. In: 2013 21st IEEE International Conference on Network Protocols (ICNP), pp. 1–6. IEEE (2013)

27. Kokila, R.T., Thamarai Selvi, S., Govindarajan, K.: DDoS detection and analysis in SDN-based environment using support vector machine classifier. In: 6th International Conference on Advanced Computing, ICoAC 2014 (2015). https://doi.org/10.1109/ICoAC.2014.7229711

28. Kolias, C., Kambourakis, G., Stavrou, A., Voas, J.: Ddos in the IoT: mirai and other botnets. Computer **50**(7), 80–84 (2017). https://doi.org/10.1109/MC.2017.201

29. Koning, R., de Graaff, B., Polevoy, G., Meijer, R., de Laat, C., Grosso, P.: Measuring the efficiency of SDN mitigations against attacks on computer infrastructures. Future Gener. Comput. Syst. **91**(1), 144–156 (2019). https://doi.org/10.1016/j.future.2018.08.011

30. Koponen, T., et al.: Onix: a distributed control platform for large-scale production networks. In: Proceedinds of the 9th USENIX Conference on Operating Systems Design and Implementation, vol. 16, no, 2, pp. 133–169 (2010). https://dl.acm.org/citation.cfm?id=279229

31. Kreutz, D., Ramos, F.M., Verissimo, P.: Towards secure and dependable software-defined networks. In: Proceedings of the Second ACM SIGCOMM Workshop on Hot Topics in Software Defined Networking - HotSDN 2013, p. 55. ACM Press, New York (2013). https://doi.org/10.1145/2491185.2491199

32. Kwon, D., et al.: A survey of deep learning-based network anomaly detection. Cluster Comput. https://doi.org/10.1007/s10586-017-1117-8

33. Lamport, L.: The part-time parliament. ACM Trans. Comput. Syst. (TOCS) **16**, 133–169 (1998). https://doi.org/10.1145/279227.279229

34. Latah, M., Toker, L.: An efficient flow-based multi-level hybrid intrusion detection system for software-defined networks. CoRR, June 2018. http://arxiv.org/abs/1806.03875

35. Latah, M., Toker, L.: Towards an efficient anomaly-based intrusion detection for software-defined networks. CoRR, March 2018. http://arxiv.org/abs/1803.06762

36. Le, A., Dinh, P., Le, H., Tran, N.C.: Flexible network-based intrusion detection and prevention system on software-defined networks. In: 2015 International Conference on Advanced Computing and Applications (ACOMP), pp. 106–111. IEEE (2015)

37. Leland, W.E., Willinger, W., Taqqu, M.S., Wilson, D.V.: On the self-similar nature of ethernet traffic. ACM SIGCOMM Comput. Commun. Rev. **25**(1), 202–213 (2004). https://doi.org/10.1145/205447.205464

38. Li, J., Zhao, Z., Li, R.: A machine learning based intrusion detection system for software defined 5G network. CoRR, July 2017. http://arxiv.org/abs/1708.04571

39. Lincoln Laboratory, Massachusetts Institute of Technology: 1999 DARPA Intrusion Detection Evaluation Dataset—MIT Lincoln Laboratory (1999). https://www.ll. mit.edu/r-d/datasets/1999-darpa-intrusion-detection-evaluation-dataset

40. Marotta, A., Carrozza, G., Avallone, S., Manetti, V.: An OpenFlow-based architecture for IaaS security. In: Proceedings of the 3rd International Conference on Application and Theory of Automation in Command and Control Systems - ATACCS 2013, p. 118. ACM Press, New York (2013). https://doi.org/10.1145/2494493.2494510

41. Mathas, C.M., et al.: Evaluation of Apache Spot's machine learning capabilities in an SDN/NFV enabled environment. In: Proceedings of the 13th International Conference on Availability, Reliability and Security - ARES 2018, pp. 1–10. ACM Press, New York (2018). https://doi.org/10.1145/3230833.3233278

42. Mckeown, N., Anderson, T., Peterson, L., Rexford, J., Shenker, S., Louis, S.: OpenFlow: enabling innovation in campus networks. ACM SIGCOMM Comput. Commun. Rev. **38**(2), 69–74 (2008). http://ccr.sigcomm.org/online/files/p69-v38n2n-mckeown.pdf

43. Jain, S., et al.: B4: Experience with a globally-deployed software defined WAN. ACM SIGCOMM Comput. Commun. Rev. **43**(4), 3–14 (2013). https://doi.org/10.1145/2534169.2486019

44. Mestres, A., et al.: Knowledge-defined networking. ACM SIGCOMM Comput. Commun. Rev. **47**(3), 4–10 (2016). https://doi.org/10.1145/3138808.3138810

45. Mohanapriya, P., Shalinie, S.M.: Restricted Boltzmann machine based detection system for DDoS attack in software defined networks. In: 2017 4th International Conference on Signal Processing, Communication and Networking, ICSCN 2017, pp. 14–19 (2017). https://doi.org/10.1109/ICSCN.2017.8085731

46. Nanda, S., Zafari, F., DeCusatis, C., Wedaa, E., Yang, B.: Predicting network attack patterns in SDN using machine learning approach. In: 2016 IEEE Conference on Network Function Virtualization and Software Defined Networks (NFV-SDN), pp. 167–172. IEEE, November 2016. https://doi.org/10.1109/NFV-SDN.2016.7919493

47. Navid, W., Bhutta, M.N.M.: Detection and mitigation of denial of service (DoS) attacks using performance aware software defined networking (SDN). In: 2017 International Conference on Information and Communication Technologies (ICICT), pp. 47–57. IEEE, December 2017. https://doi.org/10.1109/ICICT.2017.8320164

48. Neupane, R.L., et al.: Dolus. In: Proceedings of the 19th International Conference on Distributed Computing and Networking - ICDCN 2018, pp. 1–10. ACM Press, New York (2018). https://doi.org/10.1145/3154273.3154346

49. Nguyen, T.N.: The challenges in SDN/ML based network security: a survey. CoRR abs/1804-0, April 2018. https://doi.org/10.1109/CSNET.2018.8602680. http://arxiv.org/abs/1804.03539

50. Pan, J., Yang, Z.: Cybersecurity challenges and opportunities in the new "edge computing + IoT" world. In: Proceedings of the 2018 ACM International Workshop on Security in Software Defined Networks and Network Function Virtualization - SDN-NFV Sec 2018, pp. 29–32. ACM Press, New York (2018). https://doi.org/10.1145/3180465.3180470

51. Pastor, A., Mozo, A., Lopez, D.R., Folgueira, J., Kapodistria, A.: The Mouseworld, a security traffic analysis lab based on NFV/SDN. In: Proceedings of the 13th International Conference on Availability, Reliability and Security - ARES 2018, pp. 1–6. ACM Press, New York (2018). https://doi.org/10.1145/3230833.3233283

52. Prakash, A., Priyadarshini, R.: An intelligent software defined network controller for preventing distributed denial of service attack. In: 2018 Second International Conference on Inventive Communication and Computational Technologies (ICICCT), pp. 585–589. IEEE, April 2018. https://doi.org/10.1109/ICICCT.2018.8473340

53. Prasath, M.K., Perumal, B.: A meta-heuristic Bayesian network classification for intrusion detection. Int. J. Netw. Manag. **29**, e2047 (2018). https://doi.org/10.1002/nem.2047

54. Qazi, Z.A., et al.: Application-awareness in SDN. ACM SIGCOMM Comput. Commun. Rev. **43**, 487–488 (2013). https://doi.org/10.1145/2534169.2491700

55. Raj, A., Truong-Huu, T., Mohan, P.M., Gurusamy, M.: Crossfire attack detection using deep learning in software defined ITS networks. CoRR, December 2018. http://arxiv.org/abs/1812.03639

56. Rawat, D.B., Reddy, S.R.: Software defined networking architecture, security and energy efficiency: a survey. IEEE Commun. Surv. Tutor. **19**(1), 325–346 (2017). https://doi.org/10.1109/COMST.2016.2618874

57. Scott-Hayward, S., Natarajan, S., Sezer, S.: Survey of security in software defined networks. Surv. Tutor. 18(1), 623–654 (2016). https://doi.org/10.1109/COMST.2015.2474118. http://ieeexplore.ieee.org/xpls/abs_all.jsp?arnumber=7150550

58. Shin, S., Gu, G.: Attacking software-defined networks. In: Proceedings of the Second ACM SIGCOMM Workshop on Hot Topics in Software Defined Networking - HotSDN 2013, p. 165. ACM Press, New York (2013). https://doi.org/10.1145/2491185.2491220

59. Shiravi, A., Shiravi, H., Tavallaee, M., Ghorbani, A.A.: Toward developing a systematic approach to generate benchmark datasets for intrusion detection. Comput. Secur. **31**(3), 357–374 (2012). https://doi.org/10.1016/J.COSE.2011.12.012. https://www.sciencedirect.com/science/article/pii/S0167404811001672

60. Smith, R.J., Zincir-Heywood, A.N., Heywood, M.I., Jacobs, J.T.: Initiating a moving target network defense with a real-time neuro-evolutionary detector. In: Proceedings of the 2016 on Genetic and Evolutionary Computation Conference Companion - GECCO 2016 Companion, pp. 1095–1102. ACM Press, New York (2016). https://doi.org/10.1145/2908961.2931681

61. Sultana, N., Chilamkurti, N., Peng, W., Alhadad, R.: Survey on SDN based network intrusion detection system using machine learning approaches. Peer-to-Peer Netw. Appl. **12**, 1–9 (2018). https://doi.org/10.1007/s12083-017-0630-0

62. Tang, T.A., Mhamdi, L., McLernon, D., Zaidi, S.A.R., Ghogho, M.: Deep learning approach for network intrusion detection in software defined networking. In: 2016 International Conference on Wireless Networks and Mobile Communications (WINCOM), pp. 258–263. IEEE, October 2016. https://doi.org/10.1109/WINCOM.2016.7777224

63. Tang, T.A., Mhamdi, L., McLernon, D., Zaidi, S.A.R., Ghogho, M.: Deep recurrent neural network for intrusion detection in SDN-based networks. In: 2018 4th IEEE Conference on Network Softwarization and Workshops (NetSoft), pp. 202–206. IEEE, June 2018. https://doi.org/10.1109/NETSOFT.2018.8460090

64. Tantar, E., Palattella, M.R., Avanesov, T., Kantor, M., Engel, T.: Cognition: a tool for reinforcing security in software defined networks. In: Tantar, A.-A., et al. (eds.) EVOLVE - A Bridge between Probability, Set Oriented Numerics, and Evolutionary Computation V. AISC, vol. 288, pp. 61–78. Springer, Cham (2014). https://doi.org/10.1007/978-3-319-07494-8_6

65. Mininet Team: Mininet: an instant virtual network on your laptop (or other PC) - Mininet (2012). http://mininet.org/

66. University of California, Irvine: KDD Cup 1999 Data (1999). http://kdd.ics.uci.edu/databases/kddcup99/kddcup99.html

67. Wang, B., Zheng, Y., Lou, W., Hou, Y.T.: DDoS attack protection in the era of cloud computing and software-defined networking. Comput. Netw. **81**, 308–319 (2015). https://doi.org/10.1016/J.COMNET.2015.02.026. https://www.sciencedirect.com/science/article/pii/S1389128615000742

68. Wang, B., Sun, Y., Yuan, C., Xu, X.: LESLA - a smart solution for SDN-enabled mMTC E-health monitoring system. In: Proceedings of the 8th ACM MobiHoc 2018 Workshop on Pervasive Wireless Healthcare Workshop - MobileHealth 2018, pp. 1–6. ACM Press, New York (2018). https://doi.org/10.1145/3220127.3220128

69. Wang, P., Ye, F., Chen, X., Qian, Y.: Datanet: deep learning based encrypted network traffic classification in SDN home gateway. IEEE Access **6**, 55380–55391 (2018). https://doi.org/10.1109/ACCESS.2018.2872430

70. Wang, P., Chao, K.M., Lin, H.C., Lin, W.H., Lo, C.C.: An efficient flow control approach for SDN-based network threat detection and migration using support vector machine. In: Proceedings - 13th IEEE International Conference on E-Business Engineering, ICEBE 2016 - Including 12th Workshop on Service-Oriented Applications, Integration and Collaboration, SOAIC 2016, pp. 56–63 (2017). https://doi.org/10.1109/ICEBE.2016.020

71. Yan, Q., Yu, F.R., Gong, Q., Li, J.: Software-defined networking (SDN) and distributed denial of service (DDoS) attacks in cloud computing environments: a survey, some research issues, and challenges. IEEE Commun. Surv. Tutor. **18**(1), 602–622 (2016). https://doi.org/10.1109/COMST.2015.2487361

72. Yasrebi, P., Monfared, S., Bannazadeh, H., Leon-Garcia, A.: Security function virtualization in software defined infrastructure. In: 2015 IFIP/IEEE International Symposium on Integrated Network Management (IM), pp. 778–781. IEEE, May 2015. https://doi.org/10.1109/INM.2015.7140374

73. Zhang, H., Wang, Y., Chen, H., Zhao, Y., Zhang, J.: Exploring machine-learning-based control plane intrusion detection techniques in software defined optical networks. Opt. Fiber Technol. **39**, 37–42 (2017). https://doi.org/10.1016/J.YOFTE.2017.09.023. https://www-sciencedirect-com.ezproxy.unal.edu.co/science/article/pii/S1068520017303644

AIBlock - Application Intelligence and Blockchain Security

A New Proof of Work for Blockchain Based on Random Multivariate Quadratic Equations

Jintai Ding[(✉)]

University of Cincinnati, Cincinnati, USA
Jintai.Ding@gmail.com

Abstract. In this paper, we first present a theoretical analysis model on the Proof-of-Work (PoW) for cryptocurrency blockchain. Based on this analysis, we present a new type of PoW, which relies on the hardness of solving a set of random quadratic equations over the finite field GF(2). We will present the advantages of such a PoW, in particular, in terms of its impact on decentralization and the incentives involved, and therefore demonstrate that this is a new good alternative as a new type for PoW in blockchain applications.

Keywords: Proof-of-Work · Multivariate · Quadratic · NP-hard · Decentralization · Blockchain · Cryptocurrency

1 Introduction

The idea of Proof-of-work was invented in 1992 by Dwork and Naor [12] and it was initially invented to combat the spam attacks and the denial of service attacks by making such attacks economically unviable. Proof-of-Work can be defined as a protocol, which requires certain (from minimum to maximum) amount of computations in order to finish a task. The computation performed should produce something, which can be used to verify that the required amount of computations are indeed accomplished. The concept of Proof of Work has since found application. In 1997, Adam Back invented a protocol: HashCash, where Proof-of-Work is built upon a hash function. The term "PoW" was coined by M. Jakobsson and A. Juels later.

In October of 2008, Satoshi Nakamoto published his Bitcoin whitepaper [16], where Proof of Work is a key element of the Bitcoin protocol. The white paper stated that

"We propose a solution to the double-spending problem using a peer-to-peer network. The network timestamps transactions by hashing them into an ongoing chain of hash-based proof-of-work, forming a record that cannot be changed without redoing the proof-of-work."

Satoshi cited the work of HashCash. Therefore, however, the application of the PoW in Bitcoins serves a purpose very different from the original intention.

© Springer Nature Switzerland AG 2019
J. Zhou et al. (Eds.): ACNS 2019 Workshops, LNCS 11605, pp. 97–107, 2019.
https://doi.org/10.1007/978-3-030-29729-9_5

A key innovation in Bitcoin is that it uses a Proof of Work to build a competitive process called mining, which, with the help of digital signature system, help to solve three key problems:

1. prevention of the unlawful modification of the record or high cost of unlawful modification of the record;
2. synchronization of a decentralized system.
3. prevention of double spending;

In terms of our understanding, the main function of PoW is actually the property (1) and (2), since double spending is actually detected through the digital signature and therefore the stability of the blockchain due to the functionality of PoW enable us to easily solve the problem of double spending. Therefore, in bitcoin, Proof of Work systems is used to provide stability and security to an entire decentralized network, where we do not request trust on any speciifc player but the trust of the overall whole set of players. Proof of Work is used mainly to build a stable consensus mechanism, namely if there are enough mining nodes participating to perform the PoW, then the computational PoWer needed to control or attack the network becomes unattainable for any single entity. Also mining was really a great promotion tool to bring people participating in the process to therefore build a truly trustworthy decentralized system. However, as we all know, due to the appearance of the ASIC mining machines, the mining PoWer is increasingly controlled by big business players, and it does not make any sense for an ordinary user to doing mining on his or her PC on the side.

Later people invented many new PoW algorithms for various new altercoins, which we will not mention here, since none of them give a solid scientific base for their choices. In this paper, what we would like to do is to perform a complete theoretical analysis of PoW in the context of PoW for cryptocurrency, namely what are the theoretical properties we would like to have for a good PoW for a cryptocurrency. There is some initial work done before [15] in this direction but it is rather incomplete.

Then we will present our new PoW system, which is based on solving a set of random multivariate quadratic equations over GF(2), and we will show advantages of such a PoW system.

Remark 1. Here we would like to remark that the new PoW in bitcoin invented something that we humans never had before namely we can produce a document, which can be destroyed but can not be altered without incurring a tremendous cost. From this perspective, we believe PoW is much better choice than Proof of Stake (POS), since in a POS system, if someone controls the system, they can do whatever they want with essentially no cost, but in the case of PoW system for bitcoin, even if someone controls the system, if they want to do something illegitimate, to change the record, they still must pay a high prize, which is the best deterrence against the corruption of the system.

2   Theoretical Analysis of PoW in Bitcoin and Its Theoretical Model

The definition of PoW first requires certain amount of computations in order to finish a task and the computation performed should produce something, which can be used by anyone to publicly verify that the required amount of computations are indeed accomplished. From the mathematical perspective, we can view PoW as a task to solve certain mathematical problem, which can be verified by anyone. This requires clearly two properties of this mathematical problem:

1. This mathematical problem requires a certain amount computation (or a certain level of computational complexity) and no one should be able (easily) to find a new way to solve the problem that substantially reduce the computation complexity, which otherwise allow certain people to cheat. We call such a property the intrinsic hardness (**IH**) property of the problem.
2. Anyone can easily verify the solution is indeed a solution, We call this the solution public verifiability (**SPV**) property of the problem.

The IH property clearly indicates that we should use some problem that is a historicalluy very well-studied hard problem and we know very well computation complexity to solve this problem. Clearly in the case of bitcoin, Satoshi chose a very good problem, which we call the partial invertibility (PI) problem of hash functions. A simplified way to define the problem can be presented as the problem to find a string x of fixed length, such that

$$H(B, x) = (0, 0, ..., 0, *, ...*),$$

where B is the block to be mined, H is a Hash function and $*$ means values we do not care. Namely the problem is to find a preimage of any element whose first fixed number of entries are zeroes. This problem is hard to solve due to the Non-invertibility property of the hash functions, namely we can not find the preimage of a Hash function for a randomly chosen elements in the image space.

But one thing he missed, we believe, is that he did not expect that the ASIC machines can gain so much advantage compared with our ordinary PCs (scale of million), which, in some way, causes some kind of centralization of the mining PoWer being in the hands a few big miner and mining pools. However from the recent history of development of Hash function, we should also know this problem is not one of the historically well-studied problem since hash functions has a short history, and as we know the Non-invertibility hash function (like MD5) can be broken [11,18].

In the case of bitcoins, since it is a decentralized system, it means the mining problem that needs to be solved for each block in any node can be easily set up by anyone who has the information about the block, and the specific problem itself can also be easily verified. We call this the easy set-up and public verifiability (**ESUVP**) property of the problem. Since the Hash function is standardized and widely used, it clearly satisfies this property.

Also we should require that one can not easily find a block such that the problem associated with that block is substantially easier than the usual mining problem. This implies that in this family of the mathematical problems, it is computationally impossible to find a problem which is substantially easier than the hardest cases. We call such a property the homogenous hardness (**HH**) of the problem. We can see that if the problem is not HH, it is possible for someone to find an easier case and mine on such a case to gain advantage over others. In the case of bitcoin, it means that we can not easily find a meaningful B (a valid new block) such that it is easy to find x such that

$$H(B, x) = (0, 0, ..., 0, *, ...*).$$

Surely one may say that that if we find already a solution for the problem, $H(B, x) = (0, 0, ..., 0, *, ...*)$, we can repartition (B, x) to derive another solution, but this is impossible due to the fact that we want x to be of the fixed length.

In the case of bitcoin, to ensure the timing of mining of each blocks to be stable, the hardness of the problem of mining is adjusted accordingly if the mining time is substantially higher or lower than 10 min. This means we can actually adjust hardness in a controlled manner. In this case of bitcoin, the hardness is adjusted by increasing or decreasing the number of bits of zeroes, namely by increasing or decreasing the number of zeroes in the R.H.S. of

$$H(B, x) = (0, 0, ..., 0, *, ...*).$$

The hardness is basically either doubled or halfed when we increase or decrease the number of zeroes of the Hash image. In the case of Bitcoin, though we claim we have more precise adjustment, the nature of searching algorithm decides that the mining time has substantial fluctuations. Surely it will be much better to have ways to make more precisely controlled, for example, reduce hardness by a fixed percentage, however such a problem is clearly not easily to find. We will call this the difficulty adjustability (**DA**) property [15].

One more thing the bitcoin system wanted is that it should be a decentralized system that anyone can participate and make meaning contributions in the mining process. This means that the algorithm to solve the problem can be distributed independently to many independent users. We call this the independent distributability (**ID**) property of the problem. In the case of mining for bitcoin, we actually do a brutal force search on x for all $H(B, x)$, which can be easily distributed. If a problem does not have this property, then the participants will be very limited and it will not be a true decentralized system.

One more key property for PoW problems is that we should make sure that the work done for one block can not be reused substantially for another block, otherwise it will make it very hard to control the hardness since anyone can gain advantage in mining or even performing attacks. One such extreme situation is that, for any mined block, if someone can easily find a new block, which is associate to exactly the same hard problem and therefore has the same solution, then miners could reuse this problem and the solution to cheat or perform an attack by replaced published blocks with new and different blocks. In general, it is better to have the property that the mathematical problem associated with a given block to be uniquely determined, and for any given two blocks whose associated problems are totally independent and therefore no work done in one problem can be reused in another. Overall, the key is that work done for one block can not be (meaningfully) reused for another block, and this is called non-reusability (**NR**) properties [15].

Overall our conclusion is that a good PoW problem should have the following properties:

1. the intrinsic hardness (**IH**) property,
2. the solution public verifiability (**SPV**) property,
3. the easy set-up and public verifiability (**ESUVP**) property,
4. the homogenous hardness (**HH**) property,
5. the r difficulty adjustability **DA**) poperty,
6. the independent distributability (**ID**) property.
7. the non-reusability (**NR**) property.

From the above analysis, we can see that it is not easy to find such a problem, in particular, the properties IH and HH. It is clear to us many of the new PoW algorithms in altercoins are very risky choices since we know not so much about the hardness of these problems. Also it is very clear that Satoshi made a very good choice with what we knew at the time.

In [15], Kim presented a new PoW where he pointed out clearly the NR and DA property, but he did not state clearly the rest of the properties, in particular, the ID property. He presented a new POW using prime numbers, but it is not clear that his PoW satisfies all the properties we have here and we will address it in a subsequent paper.

3   A New Proof of Work Based on Random Multivariate Quadratic Equations

In this section, we propose a new idea to build a new family of PoW algorithms. The basic objects of this section are systems of quadratic polynomial equations in several variables over a finite field $\mathbb{F} = \mathbb{F}_q$ with q elements (see equation (1)).

$$p^{(1)}(x_1,\ldots,x_n) = \sum_{i=1}^{n}\sum_{j=i}^{n} p_{ij}^{(1)} \cdot x_i x_j + \sum_{i=1}^{n} p_i^{(1)} \cdot x_i + p_0^{(1)}$$

$$p^{(2)}(x_1,\ldots,x_n) = \sum_{i=1}^{n}\sum_{j=i}^{n} p_{ij}^{(2)} \cdot x_i x_j + \sum_{i=1}^{n} p_i^{(2)} \cdot x_i + p_0^{(2)}$$

$$\vdots$$

$$p^{(m)}(x_1,\ldots,x_n) = \sum_{i=1}^{n}\sum_{j=i}^{n} p_{ij}^{(m)} \cdot x_i x_j + \sum_{i=1}^{n} p_i^{(m)} \cdot x_i + p_0^{(m)}. \qquad (1)$$

A well-known hard mathematical problem is the MQ Problem:

Problem MQ: Given m quadratic polynomials $p^{(1)}(\mathbf{x}), \ldots, p^{(m)}(\mathbf{x})$ in the n variables $x_1, \ldots, x_n$ as shown in equation (1) with the coefficients of the equations uniformly and independently random chosen, find a vector $\bar{\mathbf{x}} = (\bar{x}_1, \ldots, \bar{x}_n)$ such that

$$p^{(1)}(\bar{\mathbf{x}}) = \ldots = p^{(m)}(\bar{\mathbf{x}}) = 0.$$

The MQ Problem is proven to be NP hard including quadratic polynomials over the field GF(2) [13] when m and n are of roughly the same size and this problem is known to be hard on average. In this paper we will now only concentrate on the case of GF(2).

The hardness of the MQ problem is the security foundation of the multivariate public key cryptosystems [6], a new family of post-quantum cryptosystems that can resist quantum computer attacks. Several such algorithms including Rainbow signature [9] are selected in the second round of the National Institute of Standardization of Technology post-quantum crypto standardization process [17].

The idea of using MQ problem for PoW is inspired by the work [10], where the MQ problem is used to build a Hash functions. However that Hash function has some security issues due to the collision resistance but **NOT** non-invertibility.

We will now present the construction of the new PoW algorithm based on the MQ problem.

First we will select

$$n = m + 8,$$

namely we will have 8 more variables than the number of equations.

Let us now count the number of bits of coefficients of the all the polynomials $p(i)$, which is

$$N = m \times (n(n-1)/2 + n + 1) = (n-8)(n^2/2 + n/2 + 1).$$

Suppose that we have a block B which needs to be mined. The node will then compute

$$h_i = H^i(B) = H(H(..H(B))),$$

for $i =, 1, 2 \ldots, \lceil N/256 \rceil$, if H is a 256 bit Hash function (or $i =, 1, 2 \ldots, \lceil N/512 \rceil$, if H is a 512 bit Hash function).

Then we will use h_i one by one to assign them in a fixed order to be the coefficients of the multivariate polynomials $p^{(j)}(x_1, \ldots, x_n)$, and dump any leftover bits.

Then the mining task is to find a vector

$$\bar{\mathbf{x}} = (\bar{x}_1, \ldots, \bar{x}_n)$$

such that

$$p^{(1)}(\bar{\mathbf{x}}) = \ldots = p^{(m)}(\bar{\mathbf{x}}) = 0.$$

Here we would like use the random oracle mode [2] to claim that we can view these polynomials are indeed random polynomials since we can treat a hash function as a random oracle.

Practically, the first question one may ask is that if there is indeed a solution. The answer is positive with extremely high probability.

Theorem 1. *For a MQ map with m random polynomials and $m + 8$ variables, the probability that*

$$p^{(1)}(\bar{\mathbf{x}}) = \ldots = p^{(m)}(\bar{\mathbf{x}}) = 0$$

has no solution is approximately e^{-256}, when m is big enough.

Here we will give a sketch of proof assuming the that the quadratic map is a random function. Assume that we have a map from $GF(2)^{m+8}$ to $GF(2)^m$, due to the randomness of the hash functions, we can assume that this is a random functions. Then this becomes a map from a set of size 2^{m+8} to a set of size 2^m. We would like to find out the probability that there is no vector being mapped to $(0, \ldots, 0, 0)$. It is easy to see that the probability is given as

$$(1 - 2^{-m})^{2^{m+8}} = ((1 - 2^{-m})^{2^m})^{2^8}.$$

We know that if m is big enough (like 30), we have that

$$((1 - 2^{-m})^{2^m}) \approx e^{-1}.$$

Therefore the probability will be roughly

$$((1 - 2^{-m})^{2^m})^{2^8} \approx e^{-256}.$$

This more or less guarantees that a miner can always find a solution.

Remark 2. In Bitcoin there is also no guarantee that for every block B there is a string x such that H(B——x) is of the required form. However, each miner works on a slightly different block (with different transactions) and every block in Bitcoin also contains a so called coinbase transaction (this transaction contains the Bitcoin address of the miner who wants to earn the reward money of creating the new block as well as a nonce randomly chosen by the miner.) Therefore, there exist many slightly different versions of the search problem which are worked on in parallel by the different miners. One can be relatively sure that some of these problem is actually solvable.

From the construction, it is clear that the solution public verifiability (**SPV**) property and the easy set-up and public verifiability (**ESUVP**) property are satisfied.

Due to the usage of hash functions, which can be viewed as a random oracle, to produce the problem, we can see that two different blocks can never have the same mining problems due to the collision resistance of hash functions and they should be essentially independent of each other. There the property of **NR** is satisfied.

Let us then look that the intrinsic hardness (**IH**) property, which is actually guaranteed by the NP-Hardness of the random MQ problem, and similarly we can promise the homogenous hardness (**HH**) property due to again the NP-Hardness of the random MQ problem.

Here we would like to remark that the hardness of partial inversion of a hash function in general has a much higher theoretical risk in the sense we can not actually prove the hardness of such a problem for the existing hash function, which is evident by the fact that in the past hash functions like MD5 were broken pretty badly.

Now, we would like to discuss the independent distributability (ID) property of the MQ problem. In terms of human history, more than four thousand years ago, Babylonian mathematicians could solve the problem of single variable quadratic equations. For multivariate quadratic equations, the first "smart" algorithm appeared in 1965 by Buchberger [4]. However for the problem we proposed for mining, the state of the art of the best algorithm is actually again the brutal force algorithm, which is clearly indicated in the Fukuoka MQ challenge, where there is a public challenge to find solutions for such problems (https://www.mqchallenge.org).

Due to the situation above, we also know that we can easily adjust the difficulty of the problem for each block by adjusting the number of equations where adding one more equation and one more variable means essentially doubling the hardness and reducing one equation and one variable means halfing the hardness. Therefore our PoW satisfies the **DA** property.

In addition, the new mining algorithm has other clear advantages.

1. The first one is the property we call ASIC resistance. Namely due to the following properties:
   - the simplicity of calculating the value of multivariate quadratic polynomials, which involves **very few** number of simple addition and multiplication in GF(2) after checking the first element, by this we mean that in the fast implementation of multivariate quadratic polynomial solving, we use the so called Gray code, where we search solutions in the order that each time we check if a new element is indeed a solution, it has only one bit difference from the previous element checked, to speed up tremendously the computation [3].
   More precisely, for a quadratic function $f(x_1, ..., x_n)$ over GF(2), it can be written as

   $$f(x_1, ..., x_n) = \sum_{i<j} a_{ij} x_i x_j + \sum b_i x_i + c.$$

   We will look at the case where there only 1 bit change on x_1. Assume that we already know the value of $f(x_1, ..., x_n)$ and we would like to calculate $f(x_1 + 1, ...x_n)$. We know that

$$f(x_1 + 1, ...x_n) - f(x_1 + 1, ..., x_n)$$
$$= \sum_{j>1} a_{1j}(x_1 + 1)x_j + b_1(x_1 + 1) - \sum_{j>1} a_{1j}(x_1)x_j + b_1(x_1)$$
$$= \sum_{j>1} a_{1j}x_j + b_1.$$

This means that

$$f(x_1 + 1, ...x_n) = f(x_1 + 1, ...x_n) + \sum_{j>1} a_{1j}x_j + b_1.$$

Therefore we only need to calculate

$$\sum_{j>1} a_{1j}x_j + b_1$$

to derive the value of $f(x_1 + 1, ...x_n)$ from the value of $f(x_1, ...x_n)$. Similar formula applies to the case of change of value of any variable x_i.

$$f(x_1, ., x_i, .., x_n) = f(x_1, ., x_i, .., x_n) + \sum_{i \neq j} a_{ij}x_j + b_j.$$

For such a simple calculation, GPU can achieve already great efficiency that ASIC should not be able to improve too much, while this is not the case at all in the case of PoW using hash functions.
  – the polynomials for each block actually changes all the time (almost like we use a different hash function for every block), which make it hard and costly for ASIC implementations.

we do not think that ASIC implementation can gain that much advantage compared to the usual GPUs. Surely this is based on our current technology. We believe that ASIC surely will gain advantages but it will not be more than a scale of 10 while the advantage of the case of bitcoin is about the scale of 5000. Due to the high initial cost in ASIC production, we think this design should greatly discourage the development of ASIC machines for do such a mining and therefore make it viable for small miners to do mining independently.

2. There is some work recently on attacks on PoW using quantum computers [1,8,14] due to the large key sizes of the coefficients of the MQ polynomials, the MQ-based PoW will be much harder to attack using quantum computers since it will require much more qubits for finding solutions and each time a different new set of multivariate quadratic functions has to be reloaded into the quantum system.

3. Due to the long history (thousands of years) of study of solving polynomials equations and the NP hardness of the problem, we expect that to attack our new PoW is much harder.

4. In the case of mining in bicoins, the mining is used solely to support the decentralized network. But in the MQ-based mining, the system actually rewards any progress made on solving a NP-hard problem, which is a much

more meaningful task compared to the case of hash-based mining. For example, many of the problems to attack various cryptosystems, namely many of the cryptanalysis problem, can be reduced to solving a hard seeming random quadratic systems over GF(2) and any breakthrough in this area could have tremendous effect in cryptography.

The new ideas presented in this paper is already implemented in a new cryptocurrency called ABC (www.ABCMint.org) [5,7] and it has worked very well. The research work in this paper was essentially finished in 2017. The public chain for ABC was launched on June 18, 2018. However this paper is the first publication to explain exactly the mathematics theory behind the PoW in ABC.

Surely there are many well-studied NP-hard problems, for example, the shortest vector problem (SVP) for a lattice. However if we want to use the SVP problem for mining, it is not a good choice due to the property SPV, and ESUPV. It is very hard to set up a SVP problem such that everyone can publicly verify, it is indeed a hard problem and no one can cheat in the set up process. This is why, unlike the MQ challenges, it is hard to set up pubic SVP challenges. Namely if it is indeed a random lattice, it is very hard to verify a given vector is indeed a shortest vector or not. Therefore it is actually better to use NP-complete problems, where we can easily verify the answers. A good example of a NP-complete problem is the Knapsack problem, but the Knapsack problem does not satisfy the HH property while the hardest cases of Knapsack problem is very hard but a random case is often easy to solve. To build the hard for PoW for cryptocurrency is not an easy task. From what we know by now, we believe nearly all the new PoW algorithms for altercoins needs much more careful scrutiny and they all look rather risky.

4 Conclusion

It is clear that PoW in bitcoin is different from usual PoW used in other applications and requires additional properties. We present a theoretical study of those properties required and propose a new PoW algorithms based on the MQ problem an NP-hard problem. We show the advantages of this new PoW. We hope to use this work to stimulate the research direction in PoW for cryptocurrencies.

Acknowledgment. We would like to thank Johannes Buchmann, Albrecht Petzolt, Lei Hu, Hong Xiang, Peter Ryan, Tsuyoshi Takagi, Antoine Joux, Ruben Niederhagen, Chengdong Tao, Chen-mou Cheng, Zheng Zhang, and Kurt Schmidt for useful discussions. We would like to thank the anonymous referees for useful comments. We also would like to thank the ABCMint Foundation, in particular, Jin Liu for support.

References

1. Aggarwal, D., Brennen, G.K., Lee, T., Santha, M., Tomamichel, M.: Quantum-proofing the blockchain. Quantum attacks on Bitcoin, and how to protect against them. arXiv:1710.10377 (2017)

2. Bellare, M., Rogaway, P.: Random oracles are practical: a paradigm for designing efficient protocols. In: Proceedings of the 1st ACM Conference on Computer and Communications Security, CCS 1993, pp. 62–73. ACM, New York (1993)

3. Bouillaguet, C., et al.: Fast exhaustive search for polynomial systems in $\mathbb{F}_2$. In: Mangard, S., Standaert, F.X. (eds.) CHES 2010. LNCS, vol. 6225, pp. 203–218. Springer, Heidelberg (2010). https://doi.org/10.1007/978-3-642-15031-9_14

4. Buchberger., B.: Ein Algorithmus zum Auffinden der Basiselemente des Restklassenringes nach einem nulldimensionalen Polynomideal. Ph.D. thesis, Innsbruck (1965)

5. Ding, J.: Quantum-proof blockchain. In: ETSI/IQC Quantum Safe Workshop 2018 (2018). https://www.etsi.org/events/1296-etsi-iqc-quantum-safe-workshop-2018#pane-6/

6. Ding, J., Gower, J.E., Schmidt, D.S.: Multivariate Public Key Cryptosystems. Springer, Boston (2006). https://doi.org/10.1007/978-0-387-36946-4

7. Ding, J., Liu, J.: Panel on quantum-proof blockchain. Money20/20 Hanzhou China (2018). https://www.money2020-china.com/portal/index/people/id/247.html

8. Ding, J., Ryan, P., Sarawathy, R.C.: Future of bitcoin (and blockchain) with quantum computers. Preprint of University of Cincinnati, 10.2016. Submitted to Bitcoin 2017 under Financial Cryptography 2017

9. Ding, J., Schmidt, D.: Rainbow, a new multivariable polynomial signature scheme. In: Ioannidis, J., Keromytis, A., Yung, M. (eds.) ACNS 2005. LNCS, vol. 3531, pp. 164–175. Springer, Heidelberg (2005). https://doi.org/10.1007/11496137_12

10. Ding, J., Yang, B.-Y.: Multivariates polynomials for hashing. In: Pei, D., Yung, M., Lin, D., Wu, C. (eds.) Inscrypt 2007. LNCS, vol. 4990, pp. 358–371. Springer, Heidelberg (2008). https://doi.org/10.1007/978-3-540-79499-8_28

11. Dobbertin, H.: The status of MD5 after a recent attack. CryptoBytes (2016)

12. Dwork, C., Naor, M.: Pricing via processing or combatting junk mail. In: Brickell, E.F. (ed.) CRYPTO 1992. LNCS, vol. 740, pp. 139–147. Springer, Heidelberg (1993). https://doi.org/10.1007/3-540-48071-4_10

13. Garey, M.R., Johnson, D.S.: Computers and Intractability: A Guide to the Theory of NP-Completeness. W. H. Freeman, New York (1979)

14. Gheorghiu, V., Gorbunov, S., Mosca, M., Munson, B.: Quantum-proofing the blockchain, November 2017. https://www.evolutionq.com/assets/mosca_quantum-proofing-the-blockchain_blockchain-research-institute.pdf

15. Kim, S.: Primecoin: cryptocurrency with prime number proof-of-work, March 2013. assets.ctfassets.net

16. Nakamoto, S.: Bitcoin: a peer-to-peer electronic cash system, October 2008. academia.edu

17. NIST. Post-quantum cryptograhic standardization, January 2019. https://www.nist.gov/news-events/news/2019/01/nist-reveals-26-algorithms-advancing-post-quantum-crypto-semifinals

18. Sasaki, Y., Aoki, K.: Finding preimages in full MD5 faster than exhaustive search. In: Joux, A. (ed.) EUROCRYPT 2009. LNCS, vol. 5479, pp. 134–152. Springer, Heidelberg (2009). https://doi.org/10.1007/978-3-642-01001-9_8

BSIEM-IoT: A Blockchain-Based and Distributed SIEM for the Internet of Things

Andrés Pardo Mesa[1] , Fabián Ardila Rodríguez[1] , Daniel Díaz López[1]([✉]) ,
and Félix Gómez Mármol[2]

[1] Colombian School of Engineering Julio Garavito, Bogota, Colombia
{andres.pardo-m,fabian.ardila}@mail.escuelaing.edu.co,
daniel.diaz@escuelaing.edu.co
[2] Faculty of Computer Science, University of Murcia, Murcia, Spain
felixgm@um.es

Abstract. The paper at hand proposes BSIEM-IoT, a Security Information and Event Management solution (SIEM) for the Internet of Things
(IoT) relying on blockchain to store and access security events. The security events included in the blockchain are contributed by a number of
IoT sentinels in charge of protecting a group of IoT devices. A key feature here is that the blockchain guarantees a secure registry of security
events. Additionally, the proposal permits SIEM functional components
to be assigned to different miners servers composing a resilient and distributed SIEM. Our proposal is implemented using Ethereum and validated through different use cases and experiments.

Keywords: IoT · Intrusion detection system · Blockchain · SIEM

1 Introduction

The Internet of Things (IoT) has brought uncountable benefits in a number of
diverse and relevant environments. Yet, one of its current major drawbacks lies
in the lack of security solutions to protect these systems against cyber attacks.
One approach in this regard consists in processing the security events coming
from such ecosystem and use them to prevent, detect and mitigate security incidents [2]. Security events, stemming either from IoT devices or from intermediate
security components, are collected and sent toward a centralized Security Information and Event Management (SIEM) server to detect such incidents using one
of its available modules (correlation rules, policies, statistic models).

In this regard, the integrity of the security events is critical, since an alteration of this data could induce false alarms. Likewise, availability is another
security requirement for those security events: all the security events should be
available to the SIEM modules in a timely manner, as well as resilient against

J. Zhou et al. (Eds.): ACNS 2019 Workshops, LNCS 11605, pp. 108–121, 2019.
https://doi.org/10.1007/978-3-030-29729-9_6

denial attacks. Furthermore, traceability is also a key requirement here. A comprehensive registry of all event operations should be kept and maintained to support an effective audit in case of a potential security violation.

Finally, a centralized architecture to detect intrusions in IoT ecosystems constitutes a single-point of attack and a bottle-neck that in case of failure would impact adversely all related security functions, mainly containment and recovery. Thus, resiliency becomes another requirement for the security infrastructure, so the security functions can not be interrupted.

In this paper, we present BSIEM-IoT, a blockchain-based and distributed SIEM to detect attacks against IoT devices. This proposal is built over a blockchain architecture, allowing interoperability between components of the IoT ecosystem that contribute information related to security events. Every security event is effectively protected in terms of integrity and non-repudiation due to the intrinsic features of the blockchain [7]. Further, smart contracts (SC) [8] in the blockchain guarantee a consistent behavior of the system, including the authorization of actions over the security events. BSIEM-IoT is able to consume local threat intelligence, enabling the detection of distributed attacks which can only be discovered by correlating security events coming from different sources. Moreover, our proposal connects to different external sources to get updated threat intelligence and improve the analysis of the security events within the blockchain.

The main contributions of this paper are as follows:

- A distributed SIEM proposal for IoT scenarios leveraging the benefits of a blockchain (server-less operations, integrity, non-repudiation and resiliency).
- Development of methods in a smart contract to handle blocks of security events and detect attacks from the security events available in the blockchain.
- Integration of the *External* and the *Internal Threat Intelligence* of the BSIEM-IoT to make local validations originated in smart contracts.
- The evaluation of the proposal and its features through exhaustive experiments, which in turn proved the feasibility of the solution for organizations.

2 Background

Blockchain is a decentralized P2P network where all transactions are validated by all the nodes and recorded in a distributed and immutable ledger. Consensus is the core of the blockchain technology as it guarantees the reliability of the network, and some of the existing types are presented next [11]:

- Proof of Work (PoW): A transaction is approved if at least half plus one of the nodes in the P2P network accept it.
- Proof of Stake (PoS): The node who has more wealth has greater probability to participate in the consensus and create a block.

- Proof of Importance (PoI): The nodes that can create a block are the ones with the greatest number of transactions into the network.
- Proof of Authority (PoA): Only some nodes are explicitly allowed to create new blocks and secure the blockchain.

In general, blockchain proposes two key ways to build a network [9], namely, permissioned and permissionless blockchains, being the main difference the level of governance implemented by each node. Permissionless blockchains (i.e public blockchains) allow anyone to become a node and belong to the network. Nodes on this blockchain can perform any task if they have the physical capability (e.g., mine blocks, validate transactions, etc.). In turn, permissioned blockchains (i.e. private blockchains) restrict the nodes belonging to the network and performing tasks. A relevant feature of this kind of blockchain is that it may choose the level of decentralization on the network, i.e., fully or partially decentralized.

With blockchain one can develop *Decentralized Applications* or *DApps*. To do so, a *Dapp* requires a back-end component, and in this regard, blockchains implement *smart contracts (SC)* to support any required operation by the application logic. Ethereum [1] is an open source platform to create *smart contracts*.

3   State of the Art

A number of proposals have arisen in the last years to protect IoT ecosystems. Thus for instance, [2] proposes a security architecture employing security events. Such architecture relies on a multi-relation between: (i) security events categories, providing information about the impact of an attack over a given IoT device, (ii) vulnerabilities, to explain the causes of the attack, and (iii) attack surfaces, yielding information on how the attack was conducted.

In turn, authors of [4] propose an IoT security framework for a smart home scenario. This framework applies a novel instance of blockchain by eliminating the concept of PoW and the need of coins. This work relies on a hierarchical structure that coordinates methods over the blockchain network to keep the security and privacy benefits offered by this technology. Such hierarchical structure is more suitable for the specific requirements of IoT since tasks on the network are performed in a different and adjusted manner than a common blockchain such as Bitcoin [3]. The framework proposes to manage the network and the belonging devices with the methods *store, access, monitor, genesis* and *remove.*

A blockchain-based framework to support access control in IoT is introduced in [10], implementing multiple smart contracts: (i) Access Control Contract (ACCs) to manage the authorization of users over an IoT device, (ii) Judge Contract (JC) to implement a misbehavior-judging method to facilitate the dynamic validation of the ACCs, and (iii) Register Contract (RC) to register the information of the access control and misbehavior-judging methods plus their smart contracts. When an access request arrives to the framework, different validations are done with the smart contracts before resolving such request.

In addition, [5] investigates on the applicability of a blockchain to develop the next-generation SIEM 3.0 systems, designed to detect information security incidents in a modern and fully interconnected organization network environment. This work brings the next generation of SIEM to a qualitatively new and higher level by proposing a methodology for its evaluation based on the *B method*, the most popular formal method to be used in industry projects and safety-critical system applications to allow for highly accurate expressions of the properties required by specifications and models systems in their environment.

As observed, there are already works dealing with cyber security for IoT scenarios and blockchains to tackle different IoT challenges. In particular, we found that blockchain has been applied to support IoT operations like data synchronization, communication or access control. In the paper at hand, we propose $\mathcal{B}$SIEM-IoT which, in contrast to all previous proposals, is specifically focused on the management of IoT security events. Our proposal brings the principal security features of blockchain to a regular SIEM to finally compose a security solution which is specifically focused on IoT, resilient, trust-oriented, auditable and scalable. To our best knowledge, there is no security solution applicable to IoT ecosystems holding these attributes with verifiable functionality.

4 $\mathcal{B}$SIEM-IoT

Our proposed blockchain-based and distributed SIEM for IoT, $\mathcal{B}$SIEM-IoT, validates and analyzes the compilation of security events stored in a distributed ledger of a blockchain that keeps completely safe all the information against any kind of unexpected modification. Additionally, our solution uses both internal and external threat intelligence to identify suspicious behaviors and promptly warn about an in-progress attack. Thus, $\mathcal{B}$SIEM-IoT must satisfy these goals:

- **Resilient**: In order to offer a high availability of security services, the solution should provide a go on alive capability, ensuring protection of IoT devices and attack detection, even if the SIEM gets in a hostile situation.
- **Trust-oriented**: Only trusted nodes, i.e., IoT sentinels [6], must be allowed to create transactions containing security events, avoiding data pollution.
- **Auditable**: The solution must be able to audit the block of events to identify key elements in an incident response procedure, such as identifying node(s) that issued an event or discovering causality relation between events.
- **Scalable**: The solution should be able to integrate new IoT Sentinels into the blockchain network without impacting adversely other existing nodes.

It is important to understand that a blockchain network is composed of *nodes*. While the IoT sentinels are the only ones who may create transactions in the blockchain, solely some special nodes, called miners, can receive transactions and mine (create) new blocks to be added to the blockchain. Moreover, both IoT sentinels and miners participate in the consensus algorithm.

The architecture of our proposal $\mathcal{B}$SIEM-IoT $= (\mathcal{D}, \mathcal{S}, \mathcal{M}, \mathcal{T})$ is shown in Fig. 1, encompassing the following elements: IoT devices ($\mathcal{D}$), IoT sentinels ($\mathcal{S}$), distributed SIEMs (miners $\mathcal{M}$) and external Threat Intelligence providers ($\mathcal{T}$).

4.1 IoT Devices

IoT devices ($\mathcal{D} = \{D_1, \ldots, D_{n_D}\}$) are widely deployed nowadays, including scenarios like smart homes and smart offices, amongst others. Wherever they operate, they communicate with each other and/or with other entities in the overall Internet. Due to the negative impact that a successful cyber attack would have on these (usually unprotected) devices, their communications must be secured.

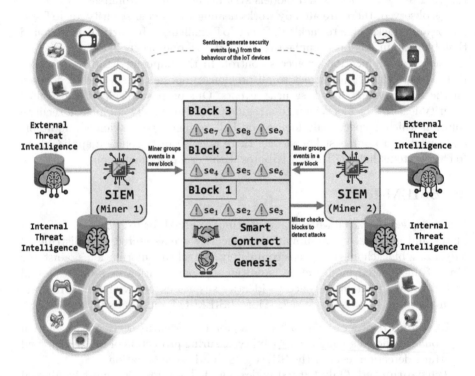

Fig. 1. Architecture of the blockchain-based and distributed SIEM, $\mathcal{B}$SIEM-IoT

4.2 IoT Sentinels

IoT sentinels ($\mathcal{S} = \{S_1, \ldots, S_{n_S}\}$) are in charge of shielding all the IoT devices in their nearby against cyber attacks. In this regard, whenever an intrusion attempt happens, the IoT sentinels generate the corresponding security event se_i and integrate it into a transaction that will subsequently be sent to the distributed SIEM (miner), who will evolve it into a block and add it to the blockchain.

Thanks to the benefits offered by the blockchain network, the sentinel here is only required to gather and keep a small portion of security events before creating an actual transaction. Thus, the sentinel just needs to run a lightweight blockchain client, turning such sentinel into a new node of the blockchain with the capacity to create transactions and to participate in a consensus validation.

The lightweight blockchain client allows the sentinel to handle smart contracts, and such smart contracts, in turn, are employed to execute several useful operations. For instance, the sentinel is able to format new security events and add them to a transaction when a given threshold of collected security events is reached. Likewise, the sentinel can also delete a specific security event from an already created transaction, so to avoid storing trash data.

4.3   Distributed SIEM (*miner*)

In contrast to the IoT sentinels, the distributed SIEM acts as the miner node in the blockchain ($\mathcal{M} = \{M_1, \ldots, M_{n_M}\}$) and must have the highest hardware features. Hence, the distributed SIEM is in charge of creating new blocks containing the transactions got from the IoT sentinels. To this end, the miner employs its computational power to solve a challenge in the blockchain network.

Moreover, the distributed SIEM will decode the information received from the lightweight blockchain client (running within the IoT sentinel) and transfer it to the external Threat Intelligence under specific formats, depending on the IoT source where the data were collected.

4.4   External Threat Intelligence

The external Threat Intelligence ($\mathcal{T}$) is provided by a third-party service analyzing malware campaigns addressed to the most prominent industries and identifying Indicators of Compromise (IoC) and Indicators of Attack (IoA) that can help another organization to detect an ongoing attack or to investigate a past attack sharing some common features with a known attack.

Intelligence information delivered by an external Threat Intelligence provider is definitely useful for BSIEM-IoT, as it may use it to analyze security events that exist in the blockchain and consequently detect IoT attacks. BSIEM-IoT is also able to incorporate this info from a third-party into its internal Threat Intelligence database, so it can be usable in the attacks detection. It is important to note that BSIEM-IoT is a distributed solution composed by a set of SIEMs, each one having different security functions and even connected to different external Threat Intelligence providers

5   Use Cases

5.1   Adding Blocks of Security Events to the Blockchain

As stated before, IoT sentinels are the only nodes in the blockchain network able to generate transactions containing security events. Yet, this action should only be granted when such devices are trustworthy enough. The novel implementation of BSIEM-IoT includes a strategic permissioned operation mode to guarantee the control and reliability of the information to be added to the blockchain.

Further, for the sake of efficiency, IoT sentinels may also group security events and include them all within the same transaction. This feature avoids creating one block for each security event, which could impact the performance of the blockchain. Thus, the **Threshold of Security Events** ($\lambda_{se} > 0$) is defined as the minimum number of events that must be grouped to create a transaction and is set previously in the configuration of the sentinels.

Finally, whenever a transaction is created by an IoT sentinel, the latter sends it to a distributed SIEM, who will in turn mine a new block with such transaction and add it to the blockchain.

5.2   Consuming the Blockchain to Detect Distributed Attacks

When $\mathcal{B}$SIEM-IoT is launched, IoT sentinels start building security events for every incident they detect. Hence, when a distributed cyber attack arises in the protected network, aiming at different IoT devices, the IoT sentinels shielding each of those victim IoT devices generate the corresponding security events.

While the IoT sentinels keep accumulating security events, they send transactions (once the threshold λ_{se} is reached) to be validated and processed by the distributed SIEM (miners). The miner processes the transaction, evolves it into a new block with all the security events and adds it to the blockchain.

In case of a distributed attack, security events related to at least two victim IoT devices are reported and added to the blockchain. If the security events are reported by two different IoT sentinels, then each of them sends its corresponding transaction to a miner. After the respective blocks are added to the blockchain, the miner consumes the security events and analyzes them using its local threat intelligence. This analysis includes the validation of security rules and policies employed to correlate security events and consequently identify distributed cyber attacks. To this end, miners can retrieve information from previous blocks stored in the blockchain. In the course of the validation process, the relevant security events are spotted and correlated to raise an alarm about the suspicious behavior.

5.3   Detecting Attacks Under Hostile Scenarios

$\mathcal{B}$SIEM-IoT is resilient against unexpected situations or even attacks aimed at the SIEM itself, without affecting its overall performance. Thus, if a miner becomes the target of a cyber attack, leading to its operational disruption, IoT sentinels would still keep generating transactions of security events. Further, the redundant and distributed additional miners, would in turn keep supporting the validation tasks needed to maintain the expected operational mode of the SIEM.

5.4   Auditing a Security Incident

Thanks to the traceability provided by the blockchain, along with the immutability of its blocks, all the information recorded in the blockchain is permanently available to be consumed in the future. Besides the security events, each block also contains data such as the address and ID of the sentinel who created the events, creation date and any information that can be useful for further analysis. Such approach allows $\mathcal{B}$SIEM-IoT to guarantee a completely auditable system.

5.5   Scaling an IoT Security Infrastructure

By leveraging the scalability properties of blockchain, BSIEM-IoT permits integrating further IoT sentinels as well as distributed SIEMs (miners) effortlessly. It is worth noting that every new node in the network (either sentinel or distributed SIEM) must be granted beforehand, prior to their actual functioning.

6   Experiments

Several preliminary experiments were conducted on the proposed solution to prove its suitability in an IoT ecosystem. Since BSIEM-IoT is composed of different elements, as shown in Fig. 1, the experiments developed in this paper have used the following infrastructure:

- IoT sentinels: Each sentinel has been deployed on a Raspberry Pi 3 model B, equipped with a quad core 1.2 GHz CPU, 1 GB RAM, 16 GB Hard Disk and OS Ubuntu Mate 16.1.
- Distributed SIEMs (miners): One SIEM (A) has been deployed in a desktop computer, equipped with a core i3 3.4 GHz x4 CPU, 5.71 GB RAM, 1.82 TB Hard Disk and OS Debian. The other SIEM (B) was deployed on a laptop Lenovo L470 equipped with Intel Core i7 7500U (2.7 GHz), 16 GB RAM, 512 GB Hard Disk and OS Debian. All SIEMs have been tested using Alienvault OSSIM[1] (Open Source SIEM) version 5.5.1.

For the ease of reading, the experiments settings are reported in Subsect. 6.1, while a significant analysis of the results is carried out in Subsect. 6.2.

6.1   Settings

The experiments were conducted by running one Ethereum [1] node on each physical component, i.e., the IoT sentinels and the SIEMs (miners). The SIEMs (miners) were able to create mined blocks thanks to their computational capabilities, whereas the IoT sentinels were only able to create transactions.

Each mined block in BSIEM-IoT is composed of a block header and a transaction. The header contains regular Ethereum header data (time stamp, difficulty, gas limit, uncles hash, gas used, among others) and the transaction includes in the **data** field the security events that were generated by IoT sentinels.

As mentioned in Sect. 5.1, BSIEM-IoT is based on a permissioned blockchain that allows only known nodes (IoT sentinels and SIEMs) to be part of the network. The consensus mechanism was the one supported currently by Ethereum, i.e. PoW; however, as Ethereum evolves, a more efficient consensus mechanism, e.g. PoS, could be used instead of PoW. PoS would reduce the time and effort that are currently required for the mining process.

---

[1] https://www.alienvault.com/products/ossim.

The reward system for $\mathcal{B}$SIEM-IoT defines its own token, which is similar to Ether, but only valid internally. In a real scenario, users interested in protecting his own IoT devices could host an IoT sentinel connected to $\mathcal{B}$SIEM-IoT to share security events. Additionally, distributed SIEMs could be hosted by different security providers at different levels like (i) Internet Service Providers (ISP), which can be interested in providing security for residential customers, (ii) National Computer Emergency Response Teams (CERTs), monitoring security incidents with a possible massive impact, or (iii) Security vendors, which can offer IoT security protection under a subscription. In this context, even if all blockchain nodes are identified, not all nodes are necessarily trusted for sharing security events. Security events are fundamental to detect and prevent attacks through the use of Threat Intelligence.

The experiments were carried out using several clients that ease the implementation of $\mathcal{B}$SIEM-IoT, namely: (i) A Remix[2] client for the IoT sentinel, which is in charge of grouping and encoding security events to be added to a new transaction, (ii) a JavaScript client for the SIEM (miner), running in the desktop computer and responsible for listening and capturing new transactions of the blockchain, in order to decode security events and make them understandable for the OSSIM server, and (iii) a JavaScript client, running in the Raspberry Pi and emulating the monitoring action that an IoT sentinel performs to generate a set of security events.

6.2   Analysis of Results

This Section offers an in-depth analysis of the outcomes from the experiments conducted over the $\mathcal{B}$SIEM-IoT. The obtained results will be organized around two kind of metrics (performance, blockchain) as shown in Table 1.

Table 1. Performance and blockchain metrics for $\mathcal{B}$SIEM-IoT

Category	Name	Description
Performance	CPU	SIEM (miner) CPU usage along an experiment time lapse
	RAM	SIEM (miner) RAM usage along an experiment time lapse
Blockchain	Number of blocks	Blocks added to the blockchain
	Gas used	Cost of carrying out an operation(s) in the Ethereum network
	Difficulty	Measure of how difficult is to generate a new block

---

[2] https://remix.ethereum.org/.

To validate the capabilities of BSIEM-IoT, two scenarios have been considered and tested:

i. Scenario 1: No critical security events (e.g. informational syslog message) are communicated from the IoT sentinel to the distributed SIEMs, which can be retained in the sentinel until reaching a Threshold of Security Events ($\lambda_{se} = 5$), and then be grouped in one transaction, until a total of 4,085 transactions is reached.

ii. Scenario 2: Critical security events (e.g. emergency syslog message) need to be communicated in a short time from the IoT sentinel to the distributed SIEMs, incorporating 1 security event per transaction, until reaching a total of 1,000 transactions.

In both cases, all the metrics have been measured over the SIEM (miner). Figures 2, 3, 4 and 5 plot the measures for each metric for both cases.

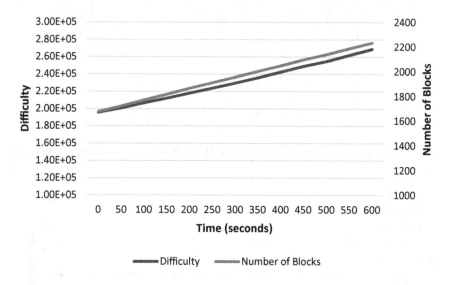

Fig. 2. Blockchain metrics for scenario 1 with 4085 transactions with 5 events per transaction

The outcomes of scenario 1 observed in Fig. 2 show how each function in a smart contract generates a gas value that defines how complex it is to execute the method in the corresponding Ethereum node. Since the first transaction was mined, 663 s elapsed until the miner created the last transaction, so, on average, each block took approximately one second to be mined.

On the other hand, the difficulty and number of blocks are directly proportional, given that every new block increases the complexity to calculate a new hash, and the difficulty considers this hash rate to be calculated. The number of blocks increases in a rate of 0.92 blocks per second, while the difficulty raises in a rate of 123.73 points of difficulty per second.

Finally, the performance metrics for the SIEM 1 in the scenario 1 (see Fig. 3) show a maximum percentage of 28.5 of used memory with some gaps where the usage of CPU is zero. When the miner is in mining process, it used practically all the CPU capability (i.e. four cores). On the other hand, the performance metrics for the SIEM B in the scenario 1 (see Fig. 3) show a constant percentage of 28.5 of used memory with some gaps where the usage of CPU is zero.

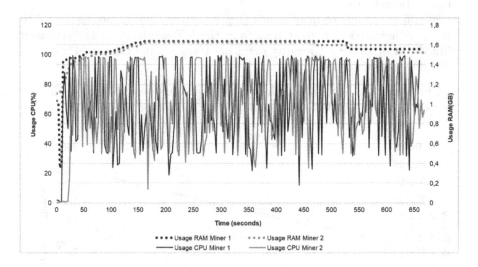

Fig. 3. Performance metrics for scenario 1 with 4085 transactions with 5 events per transaction

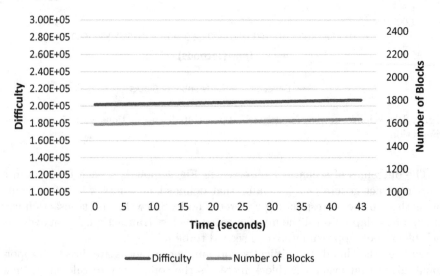

Fig. 4. Blockchain metrics for scenario 2 with 1000 transactions with 1 event per transaction

With regards to scenario 2, in Fig. 4 we observe that the time elapsed between the mining of the first and last block for this test was 43 s. In this case, where we have a greater number of transactions but lower quantity of events per transactions, every block mined took approximately 0.043 s.

After both analysis and having in mind that difficulty is adjusted periodically as function of how much hashing power has to be deployed by the network of miners, it is possible to observe that it increases with the time at different rates for each case. The above understanding let us realize that the difficulty rate is related to the block production rate which should change when more miners join the network.

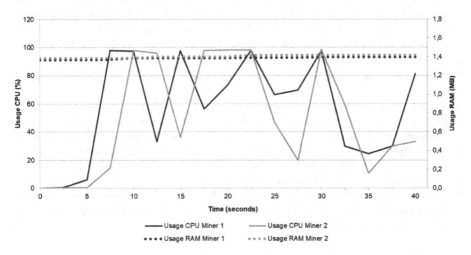

Fig. 5. Performance metrics for scenario 2 with 1000 transactions with 1 event per transaction

As for performance capabilities (see Fig. 5), in this scenario we found a similar behavior compared to the first scenario. That is to say, the miners used almost all its resources for both RAM memory and CPU usage. In this test, however, the CPU does not have gaps of zero usage, but it is rather continually in use.

As a consequence of the previously analyzed experiments, we can conclude that BSIEM-IoT yields a performance represented by a high CPU consumption (98% approx) for the CPU and a medium RAM consumption (1.4 GB approx) for the SIEMs (miners). Additionally, in both scenarios, BSIEM-IoT showcased a stable behavior with an increasing difficulty as the number of blocks grew. Last but not least, it is important to note that a block containing more events, due to the grouping made by the IoT sentinel, could require more gas since the block size is bigger in this scenario.

7 Conclusions and Future Work

By leveraging the benefits of blockchains, this paper presented $\mathcal{B}$SIEM-IoT, contributing directly to the safety of IoT ecosystems managing the security events in a strict way preserving integrity and non-repudiation. Additionally, $\mathcal{B}$SIEM-IoT offers desirable features for a sturdy security system such as resilience, trust-orientation, auditability and scalability. Experiments show that $\mathcal{B}$SIEM-IoT is able to get a desirable performance with low transaction times, which depends on the settings, being affected mainly by the Threshold of Security Events (λ_{se}) and the consensus method.

As for future works, we plan to allow new types of transactions in our solution according to the type of security event detected by the IoT sentinel, e.g. more critical security events could be added to the blockchain with a higher priority, whereas medium or low priority could be hold to be grouped. Finally, we will study the feasibility of building a new generation of IoT devices that can be blockchain-capable, qualified to report internal security events to the blockchain.

Acknowledgment. This work has been partially supported by the Escuela Colombiana de Ingeniería Julio Garavito (Colombia) through the project "Developing secure and resilient architectures for Smart Sustainable Cities" approved by the Internal Research Opening 2018 and by the project "Strengthening Governance Capacity for Smart Sustainable Cities" (grant number 2018-3538/001-001) co-funded by the Erasmus+ Programme of the European Union, as well as by a Leonardo Grant 2017 for Researchers and Cultural Creators awarded by the BBVA Foundation and by a Ramón y Cajal research contract (RYC-2015-18210) granted by the MINECO (Spain) and co-funded by the European Social Fund.

References

1. Antonopoulos, A., Wood, G.: Mastering Ethereum: Building Smart Contracts and DApps. O'Reilly Media, Sebastopol (2018)
2. Díaz López, D., et al.: Shielding IoT against cyber-attacks: an event-based approach Using SIEM. Wirel. Commun. Mob. Comput. **2018** (2018)
3. Dorri, A., Kanhere, S., Jurdak, R.: Blockchain in internet of things: challenges and solutions. CoRR abs/1608.05187 (2016)
4. Dorri, A., Kanhere, S., Jurdak, R., Gauravaram, P.: Blockchain for IoT security and privacy: the case study of a smart home. In: 2017 IEEE International Conference on Pervasive Computing and Communications Workshops (PerCom Workshops) (2017)
5. Miloslavskaya, N.: Designing blockchain-based SIEM 3.0 system. Inf. Comput. Secur. **26**(4), 491–512 (2018)
6. Nespoli, P., Useche Peláez, D., Díaz López, D., Gómez Mármol, F.: COSMOS: collaborative, seamless and adaptive sentinel for the internet of things. Sensors **19**(7), 1492 (2019)
7. Tasca, P., Tessone, C.J.: A taxonomy of blockchain technologies: principles of identification and classification. Ledger **4**, 1–39 (2019)

8. Wang, S., Ouyang, L., Yuan, Y., Ni, X., Han, X., Wang, F.: Blockchain-enabled smart contracts: architecture, applications, and future trends. IEEE Trans. Syst. Man Cybern. Syst. 1–12 (2019)

9. Wust, K., Gervais, A.: Do you need a blockchain? In: 2018 Crypto Valley Conference on Blockchain Technology (CVCBT), pp. 45–54, June 2018

10. Zhang, Y., Kasahara, S., Shen, Y., Jiang, X., Wan, J.: Smart contract-based access control for the internet of things. arXiv preprint arXiv:1802.04410 (2018)

11. Zheng, Z., Xie, S., Dai, H., Chen, X., Wang, H.: An overview of blockchain technology: architecture, consensus, and future trends. In: 2017 IEEE International Congress on Big Data (BigData Congress), pp. 557–564, June 2017

Towards Blockchained Challenge-Based Collaborative Intrusion Detection

Wenjuan Li[1], Yu Wang[2(✉)], Jin Li[2], and Man Ho Au[3]

[1] Department of Computer Science, City University of Hong Kong,
Kowloon, Hong Kong
[2] School of Computer Science, Guangzhou University, Guangzhou, China
yuwang@gzhu.edu.cn
[3] Department of Computing, The Hong Kong Polytechnic University,
Hung Hom, Hong Kong

Abstract. To protect distributed network resources and assets, collaborative intrusion detection systems/networks (CIDSs/CIDNs) have been widely deployed in various organizations with the purpose of detecting any potential threats. While such systems and networks are usually vulnerable to insider attacks, some kinds of trust mechanisms should be integrated in a real-world application. Challenge-based trust mechanisms are one promising solution, which can measure the trustworthiness of a node by sending challenges to other nodes. In the literature, challenge-based CIDNs have proven to be robust against common insider attacks, but it may still be susceptible to advanced insider attacks. How to further improve the robustness of challenge-based CIDNs remains an issue. Motivated by the recently rapid development of blockchains, in this work, we aim to combine these two and provide a blockchained challenge-based CIDN framework. Our evaluation shows that blockchain technology has the potential to enhance the robustness of challenge-based CIDNs in the aspects of trust management (i.e., enhancing the detection of insider nodes) and alarm aggregation (i.e., identifying untruthful inputs).

Keywords: Intrusion detection · Collaborative network ·
Insider attack · Blockchain technology ·
Challenge-based trust mechanism

1 Introduction

Due to the connectivity and sensing features, Internet-of-Things (IoT) has been gradually adopted by many organizations. The Gartner manager predicted that the IoT would keep delivering new opportunities for digital business innovation over the next decade, many of which can be further boosted by newly developed technologies like artificial intelligence [12]. Their report forecasts that up to 14.2 billion things will be connected by the end of 2019, and will finally reach a total of 25 billion devices by the end of 2021 [11].

© Springer Nature Switzerland AG 2019
J. Zhou et al. (Eds.): ACNS 2019 Workshops, LNCS 11605, pp. 122–139, 2019.
https://doi.org/10.1007/978-3-030-29729-9_7

The rapid growth of IoT devices brings many benefits, i.e., facilitating our daily lives, but it also becomes a major target by cyber criminals. The Symantec security report indicated that the overall volume of IoT attacks remained consistent and high in 2018 [51]. In particular, connected cameras and routers were the most infected devices - there is an increase on the infection vector. While worms and bots are still the most commonly detected IoT attacks. For example, the Mirai distributed denial of service (DDoS) worm remained an active threat and, account for 16% of the detected attacks, which was the third most common IoT threat in 2018.

To help protect the security of IoT, intrusion detection systems (IDSs) are a basic and essential security mechanism. To fit the distributed nature, collaborative intrusion detection systems/network (CIDSs/CIDNs) are often deployed in a distributed environment, which allow a set of IDS nodes to exchange required messages and understand the protected environment [54,59]. A detector could be either rule-based (signature-based) or anomaly-based. The former has to compare its stored rules with incoming events, in order to identify an attack [44,55]. The latter discovers a potential threat through identifying an anomaly between its pre-built benign profile and the current profile [45].

Insider attacks are one major threat to distributed networks and environments, hence some trust mechanisms are often implemented to protect CIDSs/CIDNs. In the literature, challenge-based trust mechanism is one promising solution, which evaluates a node's reputation by sending challenges and receiving the corresponding feedback [8]. A series of research like [8,9] has proven its effectiveness against common insider attacks; however, some studies demonstrated that such challenge-based CIDNs may still be susceptible to advanced attacks [23–25,27]. For instance, the Passive Message Fingerprint Attack (PMFA) [23] enables suspicious nodes to cooperate in identifying normal messages and remain their reputation without being detected. Thus, there is a great need to design a more robust challenge-based CIDNs to ensure its detection effectiveness. Below are three desirable attributes for a new CIDN framework.

– To avoid the issue of a single point of failure (SPOF), the CIDN framework should not rely mainly on a centralized server.
– The CIDN framework should provide a robust trust management process, which can evaluate the trustworthiness of nodes in an accurate way.
– The CIDN framework should be able to identify malicious inputs, which are even from some trusted nodes.

Recently, blockchain technology has become quite popular encouraged by the success of cryptocurrency Bitcoin. The Gemalto report [10] indicates that the adoption of blockchains has doubled from 9% to 19% in the early 2019, and this trend is likely to continue in the next year and beyond. They also described a survey that up to 23% of respondents believed that blockchain technology would be an ideal solution to use for securing IoT devices, and 91% of organisations are likely to consider it in the future. For instance, Amazon announced its new managed service, Amazon Managed Blockchain, which allows users to set up and configure a scalable blockchain network with just a few clicks [2]. With a huge number of devices, blockchains can increasingly be used to monitor and record those communications and transactions in an IoT environment [29].

Currently, blockchains have been applied into many domains like IoT [28,48], transportation [17,22], and energy [47]. The strong encryption used to secure blockchains can greatly increase the difficulty for cyber criminals to brute-force their way into private and sensitive environments. Due to these merits, some research has started trying to combine blockchains with CIDSs/CIDNs. An initial blockchain-based framework was proposed by Alexopoulos et al. [1], aiming to protect the alarm exchange among the collaborating nodes. They regarded raw alarms generated by the monitors are stored as transactions in a blockchain, replicated among the participating nodes of peer-to-peer network. While they did not show any experimental implementation or results. Tug et al. [52] introduced CBSigIDS, a framework of collaborative blockchained signature-based IDSs, by incrementally sharing and building a trusted signature database via blockchains in a CIDN network. They mainly targeted the combination of blockchains with signature-based IDSs, but remained anomaly-based detection as future work. On the other hand, a blockchain-based framework called CIoTA was proposed by Golomb *et al.* [13], which focused solely on anomaly detection via updating a trusted detection model.

Contributions. Though some studies have discussed the intersection between CIDSs and blockchains, to the best of our knowledge, most existing work was initialized at the high level, without specifying a concrete CIDS/CIDN. In addition, there is no work focusing on a specific trust-based detection system. To make up this gap, in this work, we focus on the challenge-based trust mechanism, and develop a blockchained challenge-based CIDN framework. Our contributions can be summarized as below.

- To combine the blockchain technology with a concrete type of trust-based CIDN, we propose a blockchained challenge-based CIDN framework, which can be workable under both signature-based and anomaly-based detection. In particular, blockchains can be served as an additional layer to provide the flexibility in practical deployment.
- Under our framework, we show how to use blockchains to enhance the robustness of trust management against attacks, as well as protect the alarm aggregation process from malicious inputs. The enhancement is valid for both signature-based and anomaly-based detection.
- In the evaluation, we exploit the performance of our framework in the aspects of trust computation and alarm aggregation. Our results demonstrate that our framework can become more robust via the implementation of blockchains, i.e., identifying malicious nodes and untruthful inputs.

Paper Organization. Section 2 introduces research studies on collaborative intrusion detection and the background of blockchains. Section 3 describes our framework of blockchained challenge-based CIDNs that can be suitable for both signature-based and anomaly-based detection. We show how to use blockchains to enhance the trust management and alarm aggregation. Section 4 shows our experimental settings and analyzes the collected results. We discuss some challenges in Sect. 5 and conclude the work in Sect. 6.

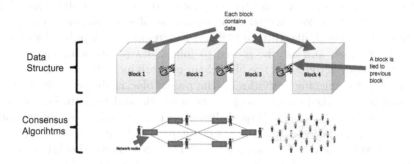

Fig. 1. The high-level review of blockchains.

2 Background and Related Work

In this section, we introduce the background of blockchain technology and review research studies on distributed detection systems, collaborative intrusion detection and blockchain-based detection.

2.1 Background of Blockchains

The original purpose of blockchains is to make payments between entities without a trust relationship and build a temper-resistant blockchain. Cryptocurrencies like Bitcoin have proven to be a phenomenal success. The underlying blockchain technique, which is an ingenious combination of multiple technologies such as peer-to-peer network, consensus protocol over a distributed network, cryptographic schemes, distributed database, smart contract and game theory, provides a decentralized way to build trust in our social and economic activities, and thus holds a huge promise to change the future of financial transactions, and even our way of computation and collaboration. As one of the hottest topics in the fields of IT and Fintech, blockchain has drawn much attention from researchers, as well as IT and FinTech industry. So far, both research community and industry community have made significant progresses in blockchain technologies and applications.

A blockchain node often maintains a list of records (known as blocks), which are organized in a chronological order based on discrete time stamps [60]. A block is typically comprised of a payload, a timestamp and a cryptographic hash value. The first block is called genesis block, and the node behind can connect to the previous one via a hash value. New blocks are added in a sequential manner with the next block containing a hash of the previous block. A new block can be generated once the previous block enters in the blockchain. The big feature of a block is that the recorded data in any block could not be modified without the alteration of all subsequent blocks [38]. The high-level review of blockchains is depicted in Fig. 1.

A blockchain can be generally classified into two categories: public blockchain and permissioned blockchains [60]. The former enables anyone to join and contribute to the network like Bitcoin [39] and Ethereum [58]. A public blockchain is completely open and anyone is free to join & leave. Everyone can participate in the major activities of the blockchain network including reading, writing and auditing the ongoing activities on the public blockchain network. The latter allows only verified entities to join the network, and perform only certain activities on the network like Hyperledger [15]. For example, Such blockchains would grant special permissions to each participant to have permissions to read, access and write pre-defined information on the blockchains. Blockchain nodes can make a decision-making process via consensus algorithms. There are some requirements for consensus algorithms in blockchains. For instance, the algorithm should collect all the agreements from chain nodes. Each node should aim at a better agreement to fit a whole interest.

There are may related studies focused on consensus mechanism. Badertscher et al. [3] put forth the first global universally composable (GUC) treatment of PoS-based blockchains in a setting that captures arbitrary numbers of parties that may not be fully operational (i.e., dynamic availability, which naturally captures decentralized environments within which real-world deployed blockchain protocols are assumed to operate). They proposed a new PoS-based protocol called "Ouroboros Genesis" which enables new or offline parties to safely (re-) join and bootstrap their blockchain from the genesis block without any trusted advice (such as checkpoints) or assumptions regarding past availability. With the model allowing adversarial scheduling of messages in a network with delays and captures the dynamic availability of participants in the worst case, the authors proved the GUC security of Ouroboros Genesis against a fully adaptive adversary controlling less than half of the total stake. Kiffer et al. [16] developed a simple Markov-chain based method for analyzing consistency properties of blockchain protocols. This method could be used to address a number of basic questions about consistency of blockchains such as providing a tighter guarantee on the consistency property of Nakamoto's protocol, analyzing a family of delaying attacks and extending them to other protocols, giving the first rigorous consistency analysis of GHOST, and so on. Wan et al. [56] presented a novel hybrid consensus protocol named Goshawk, in which a two-layer chain structure with two-level PoW mining strategy and a ticket-voting mechanism are elaborately combined. They showed that Goshawk is the first blockchain protocol with three key properties such as high efficiency, strong robustness against the 51% attack.

Pass et al. [42] proposed a new paradigm called Thunderella for achieving state machine replication by combining a fast, asynchronous path with a (slow) synchronous "fall-back" path. With this paradigm, they provided a new resilient blockchain protocol (for the permissionless setting) assuming only that a majority of the computing power is controlled by honest players, and optimistically, transactions could be confirmed as fast as the actual message delay in the network if 3/4 of the computing power is controlled by honest players, and a special player called the accelerator is honest. Daian et al. [4] presented a provably secure proof-of-stake protocol called Snow White. As a matter of fact, Snow White was publicly released in 2016. It provides a formal, end-to-end proof of a proof-of-stake system in a truly decentralized, open-participation network. The authors identified a core "permissioned" consensus protocol suitable for proof-of-stake, and proposed a robust committee re-election mechanism such that as stake switches hands, the consensus committee can evolve in a timely manner and always reflect the most recent stake distribution. They also introduced a formal treatment of costless simulation issue and gave both upper- and lower-bounds that characterize exactly what setup assumptions are needed to resist costless simulation attacks.

2.2 Related Work

In real-world applications, a separate IDS often has no information about its deployed and protected environment, opening a chance for attackers and cyber-criminals. Due to the lack of contextual information, it becomes very hard for an IDS to figure out complicated attacks. Focus on this issue, there is a great need for building a distributed system or collaborative network to enhance the detection performance [59].

Distributed Systems. Distributed systems have been widely used in various domains over many years. For example, Prras et al. [43] introduced EMERALD (Event Monitoring Enabling Responses to Anomalous Live Disturbances) in 1997, which aimed to monitor malicious behaviors across different layers in a large network. It can model distributed high-volume events and correlate them using traditional IDS techniques. Snapp et al. [46] presented a distributed Intrusion Detection System (DIDS), which could improve the monitoring process with data reduction method and centralized data analysis. Then, COSSACK system [41] was developed to reduce the impact of DDoS attack, which could work without the support and inputs from humans, i.e., it could generate rules and signatures in an automatic way. Then, DOMINO (Distributed Overlay for Monitoring InterNet Outbreaks) [61] was proposed, aiming to enhance the collaboration process among different nodes. They particularly used an overlay design to achieve a heterogeneous, scalable, and robust mechanism. PIER [14] was an Internet-scale query engine and a kind of querying-based system. It could help distribute dataflows and queries in a better way.

Collaborative Intrusion Detection. A collaborative system encourages an IDS node to collect and exchange information with other nodes. Li et al. [18] found that most distributed intrusion detection architectures could not be scalable under different communication mechanisms. Thus, they proposed a distributed detection system by means of a decentralized routing infrastructure. However, one big limitation is that all nodes in their approach should be intra trusted. This may lead to insider attacks, which are one common threat for various distributed systems and collaborative networks.

To protect distributed/collaborative systems against insider attacks, it is very important to design suitable trust mechanisms to measure the reputation in such systems and networks. As an example, an overlay IDS was proposed by Duma *et al.* [5], which could identify insider attacks. It consists of a trust-aware engine for correlating alarms and an adaptive trust mechanism for handling trust. Then Tuan [53] applied game theory to help enhance the detection performance in a P2P network. They found that if a trust system was not incentive compatible, the more numbers of nodes in the system, the less likely that a malicious node would be identified.

Fung *et al.* [8] proposed a kind of challenge-based CIDNs, which could evaluate the trustworthiness of an IDS node based on the received answers to the challenges. They first proposed a collaboration framework for host-based IDSs with a forgetting factor, which can emphasize on the recent behavior of a node. To enhance such challenge mechanisms, Li *et al.* [19] claimed that IDS nodes may have different sensitivity levels in identifying particular intrusions. Then they proposed a concept of *intrusion sensitivity (IS)* that measures the detection sensitivity of an IDS for a particular intrusion. They also designed an *intrusion sensitivity-based trust management model* [20] that could automatically allocate the values by using machine learning classifiers like KNN classifier [34]. They also performed a study to investigated the effect of intrusion sensitivity on detecting pollution attacks, where a set of malicious nodes collaborate to affect alert rankings by offering untruthful information [21]. They indicated that *IS* can help decrease the reputation of malicious nodes quickly. Other related work regarding how to improve the performance of intrusion detection can refer to [6,7,30–33,36,37,57].

Blockchain-Based Intrusion Detection. The application of blockchains in the field of intrusion detection has been studied, but it is still an emerging topic. Alexopoulos *et al.* [1] described a framework to show how to combine a blockchain with a CIDS. They considered a set of raw alarms produced by each IDS as transactions in a blockchain. Then, all collaborating nodes could use a consensus protocol to ensure the transaction validity before delivering them in a block. This can make sure that the stored alarms are tamper resistant in the blockchain. The major limitation is that they did not provide any results or implementation detail.

Then Meng *et al.* [38] provided the first review regarding the intersection of blockchains and intrusion detection, and introduced the potential application of such combination. They indicated that blockchains can help enhance an IDS in the aspects of data sharing, trust computation and alarm exchange. For anomaly detection, Golomb *et al.* [13] described a framework called CIoTA, which could apply blockchains to perform anomaly detection in a distributed manner for IoT devices. By contrast, Li et al. [26] demonstrated how to use blockchains to enhance the performance of collaborative signature-based IDSs via building a verifiable rule database. On the other hand, some studies investigated how an IDS can help protect blockchain applications. Steichen *et al.* [50] introduced an OpenFlow-based firewall named ChainGuard, which could help protect blockchain-based SDN and identify malicious traffic and behavior in the network.

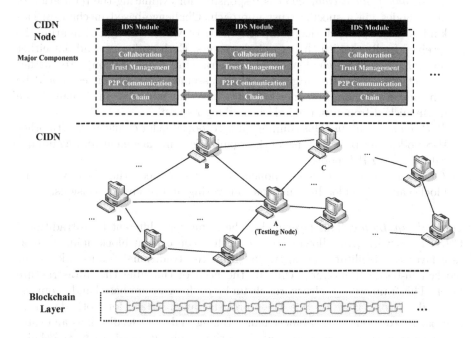

Fig. 2. Blockchained challenge-based CIDN framework: a high-level review.

3   Our Proposed Framework

As discussed above, there are already some studies investigating the intersection of collaborative intrusion detection and blockchains. While most of them (e.g., [1]) focused on a generic CIDS without considering a particular trust mechanism. In practice, the implementation of blockchains may depend on the specific types of trust mechanisms. In this section, we propose a blockchain-based framework for challenge-based CIDNs particularly.

3.1   Framework Design

Figure 2 shows the high-level framework of blockchained challenge-based CIDNs. Under the CIDN, an IDS module is a basic component. There are some more major components: *collaboration component, trust management component, P2P communication,* and *chain component.*

- *Collaboration component* is mainly responsible for assisting a node in computing the trust values of another node by sending out *normal requests* or *challenges*, and receiving the relevant *feedback*. This component can help a tested node deliver its feedback when receiving a request or challenge. For instance, Fig. 1 shows that when node A sends a *request* or *challenge* to node B, it can receive relevant feedback.
- *Trust management component* is responsible for evaluating the reputation of other nodes via a specific trust approach. Challenge-based mechanism is a kind of trust approach that computes the trust values through comparing the received feedback with the expected answers. Each node can send out either normal requests or challenges for alert ranking (consultation). To further protect challenges, the original work [8] assumed that challenges should be sent out in a random manner and in a way that makes them difficult to be distinguished from a normal alarm ranking request.
- *P2P communication.* This component is responsible for connecting with other IDS nodes and providing network organization, management and communication among IDS nodes.
- *Chain component.* This component aims to connect the node with the blockchain, i.e., uploading information, voting and receiving decisions.

Blockchain Layer. This layer makes the framework different from traditional CIDN frameworks, by allowing to establish a consortium blockchain. A separate layer can facilitate the migration from the traditional framework to our blockchain-based framework, without the need of changing the infrastructure much. This framework is also workable under both signature-based and anomaly-based detection. That is, this layer provides an interface for both detection approaches to connect with blockchains. Taking malicious feedback as an example, each chain node can check and share the suspicious feedback to the chain, and other chain nodes can help verify the feedback. This can help either build a trusted rule database [26] or enhanced profile [13].

In such network, every IDS node can select its own partners according to defined policies, and maintain a list called *partner list*. When a node wants to join the CIDN, it first has to apply and get a unique proof of identity (e.g., a public and a private key pair) via a trusted certificate authority (*CA*). As depicted in Fig. 1, if node B asks for joining the network, it has to send a request to a CIDN node, say node A. Then, node A makes a decision and sends back an initial *partner list*, if node C is accepted. A CIDN node can typically send two types of messages: *challenge* and *normal request*.

- A *challenge* mainly contains a set of IDS alarms, where a testing node can send these alarms to the tested nodes for labeling alarm severity. Because the testing node knows the severity of these alarms in advance, it can judge and compute the satisfaction level for the tested node, based on the received feedback.
- A *normal request* is sent by a node for alarm aggregation, which is an important feature of collaborative networks in improving the detection performance

of a single detector. The aggregation process usually only considers the feedback from highly trusted nodes. As a response, an IDS node should send back alarm ranking information as their feedback.

3.2 Trust Management

Node Expertise. In this work, we consider three expertise levels for an IDS node as low (0.1), medium (0.5) and high (0.95). The expertise of an IDS can refer to a beta function described as below:

$$f(p'|\alpha, \beta) = \frac{1}{B(\alpha, \beta)} p'^{\alpha-1}(1 - p')^{\beta-1}$$
$$B(\alpha, \beta) = \int_0^1 t^{\alpha-1}(1 - t)^{\beta-1} dt \tag{1}$$

where $p'(\in [0, 1])$ is the probability of intrusion examined by the IDS. $f(p'|\alpha, \beta)$ means the probability that a node with expertise level l responses with a value of p' to an intrusion examination of difficulty level $d(\in [0, 1])$. A higher value of l means a higher probability of correctly identifying an intrusion while a higher value of d means that an intrusion is more difficult to detect. In particular, α and β can be defined as [9]:

$$\alpha = 1 + \frac{l(1 - d)}{d(1 - l)} r$$
$$\beta = 1 + \frac{l(1 - d)}{d(1 - l)}(1 - r) \tag{2}$$

where $r \in \{0, 1\}$ is the expected result of detection. For a fixed difficulty level, the node with higher level of expertise can achieve higher probability of correctly detecting an intrusion. For example, a node with expertise level of 1 can accurately identify an intrusion with guarantee if the difficulty level is 0.

Node Trust Evaluation. To measure the reputation of a target node, a testing node can deliver *challenges* via a random generation process. Then the testing node can calculate a score to indicate the satisfaction. According to [8], we can evaluate the reputation of a node i according to node j as follows:

$$T_i^j = (w_s \frac{\sum_{k=0}^n F_k^{j,i} \lambda^{tk}}{\sum_{k=0}^n \lambda^{tk}} - T_s)(1 - x)^d + T_s \tag{3}$$

where $F_k^{j,i} \in [0, 1]$ is the score of the received feedback k and n is the total number of feedback. λ is a *forgetting factor* that assigns less weight to older feedback response. w_s is a *significant weight* depending on the total number of received feedback, if there is only a few feedback under a certain minimum m, then $w_s = \frac{\sum_{k=0}^n \lambda^{tk}}{m}$, otherwise $w_s = 1$. x is the percentage of "don't know" answers during a period (e.g., from $t0$ to tn). d is a positive incentive parameter to control the severity of punishment to "don't know" replies. More details about equation derivation can be referred to [8].

Satisfaction Evaluation. Intuitively, satisfaction can be measured between an expected feedback ($e \in [0,1]$) and an actual received feedback ($r \in [0,1]$). In addition, we can construct a function F ($\in [0,1]$) to derive the satisfaction score as follows [8,9]:

$$F = 1 - (\frac{e-r}{max(c_1 e, 1-e)})^{c_2} \quad e > r \tag{4}$$

$$F = 1 - (\frac{c_1(r-e)}{max(c_1 e, 1-e)})^{c_2} \quad e \le r \tag{5}$$

where c_1 controls the degree of penalty for wrong estimates and c_2 controls satisfaction sensitivity. A larger c_2 means more sensitive. In this work, we set $c_1 = 1.5$ and $c_2 = 1$ based on the simulation in [9].

In Combination with Blockchains. The blockchained challenge-based CIDN can be treated as a consortium blockchain, as each node should be verified by a CA and get their key pair. It is a key to enhance the robustness of trust computation by measuring the received feedback. In this case, we can submit the received feedback to the chain for verification. If it is not passed, then the feedback can be considered as a suspicious one.

3.3   Alarm Aggregation

Alarm aggregation is a critical process, which can help such collaborative systems make a decision. Intuitively, a node performing the process can request the alarm rankings from other trusted nodes in its *partner list*. For instance, node j can aggregate the feedback $R_j(a)$ from others, and make a decision, e.g., the aggregated ranking of alert a, by using a weighted majority method as below.

$$R_j(a) = \frac{\sum_{T \ge r} T_i^j D_i^j R_i(a)}{\sum_{T \ge r} T_i^j D_i^j} \tag{6}$$

where $R_i(a)(\in [0,1])$ indicates the aggregated ranking of alert a by node i, r means a trust threshold that node j only accepts the alarm ranking from those nodes whose reputation is higher than this threshold. $T_i^j(\in [0,1])$ indicates the reputation of node i according to node j. $D_i^j(\in [0,1])$ describes how many *hops* between these two nodes.

In Combination with Blockchains. The alarm aggregation is a critical process in CIDNs, in which an IDS node decides whether there is an intrusion or not. In real-world applications, some malicious nodes may have high reputation at first (e.g., betrayal nodes) and can send untruthful alarm feedback. To avoid the negative impact, the blockchained challenge-based CIDN can submit the received alarm ranking to the chain for validation. If any suspicious clues are found, then the received alarm feedback can be discarded.

4   Evaluation: A Case Study

In this section, as a first study, our purpose is to evaluate the initial performance of our framework in a simulated environment, where malicious nodes could perform an advanced collusion attack, called random poisoning attack [35]. It enables malicious nodes making untruthful feedback with a possibility. In practice, the possibility can be tuned according to the requirements from different environments and networks. The simulated environment contains 50 nodes that are randomly distributed in a 12×12 grid region. We deployed an IDS, e.g., Snort [49] and Zeek [62] in each node, and all IDS nodes can find their own partners after communicating with others within a time period. The consortium blockchain was deployed in a mid-end computer with Intel(R) Core (TM)i6, CPU 2.5 GHz with 100 GB storage.

To evaluate the trustworthiness of partner nodes, each node can send out challenges randomly to its partners with an average rate of ε. There are two levels of request frequency: ε_l and ε_h. For the nodes that have a unclear trust value around the threshold, the frequency should be set as high ε_h. The detailed parameters are shown in Table 1. All the settings are maintained similar to relevant work [8,20,24].

Table 1. Parameter settings in the experiment.

Parameters	Value	Description
λ	0.9	Forgetting factor
m	10	Lower limit of received feedback
d	0.3	Severity of punishment
ε_l	10/day	Low request frequency
ε_h	20/day	High request frequency
r	0.8	Trust threshold
T_s	0.5	Trust value for newcomers

Trust Evaluation Under Attack. We randomly selected three expert nodes to perform the random poisoning attack. In particular, a malicious node under random poisoning attack enjoys a possibility of $1/2$ in sending out malicious feedback. Figure 3 depicts the reputation of malicious nodes under both traditional framework and our blockchain-based framework.

- It is observed that the trustworthiness of malicious nodes could be reduced faster under our framework than that under the traditional framework. This is because traditional framework cannot identify all malicious feedback nodes as the malicious nodes only behave untruthfully with a possibility.
- By contrast, our framework leverages the application of blockchains and each feedback could be verified by all chain nodes. This can greatly increase the successful rate of detecting malicious feedback. Thus, our framework can decrease the reputation of malicious nodes in a fast manner.

Alarm Aggregation Under Attack. Similarly, we also selected three expert nodes randomly to deliver false alarm rankings to a node that performs alarm aggregation. We mainly consider a false negative (FN) rate and a false positive (FP) rate. Figure 4 shows the errors of alarm aggregation under both traditional framework and our framework.

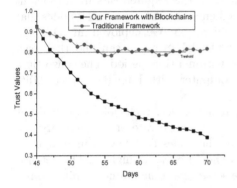

Fig. 3. The trust values of malicious nodes: under traditional and our framework.

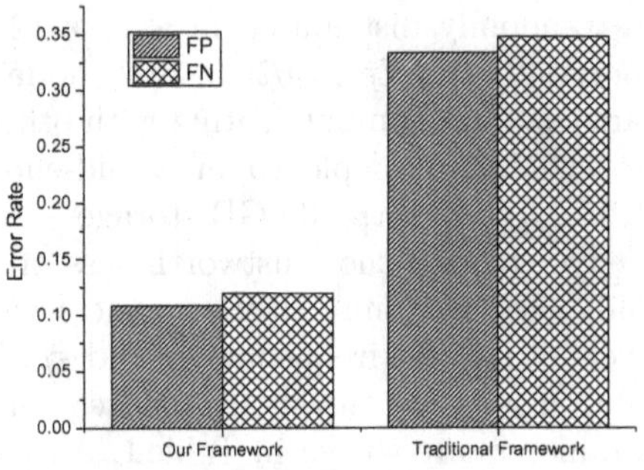

Fig. 4. Classification errors during the alarm aggregation process.

- It is found that the errors under the traditional framework are generally high with $FN = 33.3\%$ and $FP = 34.8\%$. This is because the traditional framework cannot identify malicious nodes efficiently, e.g., under the random poisoning attack. Therefore, these malicious nodes could still make a negative impact on the alarm aggregation.
- In the comparison, our framework could reduce the error rates significantly, i.e., with $FN = 10.8\%$ and $FP = 11.9\%$. There are two major reasons. One is that our framework can help identify malicious nodes in a quick manner, e.g., under the random poisoning attack. Also, in our framework, the received alarm rankings can be submitted to the chain for verification, and it is easier to detect untruthful inputs, even from trusted nodes, i.e., betrayal nodes.

Overall, our study indicates that our framework can enhance the robustness of challenge-based CIDNs in the aspects of both trust management and alarm aggregation, through integrating with blockchains.

5   Discussion and Challenges

Though blockchain technology can bring a lot of benefits, it is still at a developing stage, which may suffer many challenges from both inside and outside [38].

- *Energy and cost.* The computational power is a concern for blockchain applications in real-world scenarios. For example, Proof of Work (PoW) may require huge amounts of energy while doing bitcoin mining, where the electricity consumption could rise to 7.7 GW by the end of 2018, which is almost half a percent of the world's electricity consumption.
- *Security and privacy.* Though Bitcoin has been widely adopted, it does not mean that it is safe. There are existing some types of attacks. Taking eclipse attack as an example, as the chain nodes have to keep constant communication to compare data, an attacker can fool it into accepting false data if he/she has successfully compromised that node [40]. This results in wasting network resources or accepting fake transactions. There is a need to enhance the security of blockchain itself.
- *Complexity and speed.* Blokchain is a complex system that is hard to be established from scratch. A single mistake may cause the whole system to be compromised. Due to the complexity, it also suffers data storage and transaction speed issues. As a study, we only tried a proof-of-concept chain to investigate the performance. It is an important topic to exploit the practical performance when the blockchain runs for a while.
- *Blockchain size.* In the beginning of a blockchain, the node number may be in a small scale, which makes it vulnerable to many attacks during the growth. For instance, assume there are only 30 nodes, if a single entity successfully controls just or more than 51% of the blockchain nodes, then it has a high probability to control the whole outputs.

6   Conclusion

Challenge-based Collaborative intrusion detection provides a promising solution to safeguard assets from being compromised; however, it may still be vulnerable to advanced attacks in practical deployment. Motivated by the fast development of blockchains, in this work, we propose a blockchained challenge-based CIDN framework by leveraging the benefits offered by the blockchain technology. Our framework enables nodes to form a consortium chain and improve the robustness of challenge-based CIDNs. In the evaluation, our results demonstrate that our framework can enhance the robustness in the aspects of trust management by detecting advanced malicious nodes, and alarm aggregation through identifying untruthful inputs and reducing error rates.

Acknowledgments. This work was funded by the National Natural Science Foundation of China (NSFC) Grant No. 61772148, 61802080 and 61802077.

References

1. Alexopoulos, N., Vasilomanolakis, E., Ivánkó, N.R., Mühlhäuser, M.: Towards blockchain-based collaborative intrusion detection systems. In: D'Agostino, G., Scala, A. (eds.) CRITIS 2017. LNCS, vol. 10707, pp. 107–118. Springer, Cham (2017). https://doi.org/10.1007/978-3-319-99843-5_10

2. Amazon Managed Blockchain: Easily create and manage scalable blockchain networks. https://aws.amazon.com/managed-blockchain/. Accessed 10 Apr 2019
3. Badertscher, C., Gazi, P., Kiayias, A., Russell, A., Zikas, V.: Ouroboros genesis: composable proof-of-stake blockchains with dynamic availability. In: Proceedings of ACM Conference on Computer and Communications Security (CCS), pp. 913–930 (2018)
4. Daian, P., Pass, R., Shi, E.: Snow white: robustly reconfigurable consensus and applications to provably secure proofs of stake. In: Financial Cryptography and Data Security (FC) (2019)
5. Duma, C., Karresand, M., Shahmehri, N., Caronni, G.: A trust-aware, P2P-based overlay for intrusion detection. In: DEXA Workshop, pp. 692–697 (2006)
6. Fadlullah, Z.M., Taleb, T., Vasilakos, A.V., Guizani, M., Kato, N.: DTRAB: combating against attacks on encrypted protocols through traffic-feature analysis. IEEE/ACM Trans. Netw. **18**(4), 1234–1247 (2010)
7. Friedberg, I., Skopik, F., Settanni, G., Fiedler, R.: Combating advanced persistent threats: from network event correlation to incident detection. Comput. Secur. **48**, 35–47 (2015)
8. Fung, C.J., Baysal, O., Zhang, J., Aib, I., Boutaba, R.: Trust management for host-based collaborative intrusion detection. In: De Turck, F., Kellerer, W., Kormentzas, G. (eds.) DSOM 2008. LNCS, vol. 5273, pp. 109–122. Springer, Heidelberg (2008). https://doi.org/10.1007/978-3-540-87353-2_9
9. Fung, C.J., Zhu, Q., Boutaba, R., Basar, T.: Bayesian decision aggregation in collaborative intrusion detection networks. In: NOMS, pp. 349–356 (2010)
10. Almost half of companies still can't detect IoT device breaches, reveals Gemalto study. https://www.gemalto.com/press/Pages/Almost-half-of-companies-still-can-t-detect-IoT-device-breaches-reveals-Gemalto-study.aspx. Accessed 10 Apr 2019
11. Leading the IoT: Gartner Insights on How to Lead in a Connected World. https://www.gartner.com/imagesrv/books/iot/iotEbook_digital.pdf. Accessed 22 Mar 2019
12. Gartner Identifies Top 10 Strategic IoT Technologies and Trends. https://www.gartner.com/en/newsroom/press-releases/2018-11-07-gartner-identifies-top-10-strategic-iot-technologies-and-trends. Accessed 22 Mar 2019
13. Golomb, T., Mirsky, Y., Elovici, Y.: CIoTA: Collaborative IoT Anomaly detection via blockchain. In: Proceedings of Workshop on Decentralized IoT Security and Standards (DISS), pp. 1–6 (2018)
14. Huebsch, R., et al.: The architecture of PIER: an internet-scale query processor. In: Proceedings of the 2005 Conference on Innovative Data Systems Research (CIDR), pp. 28–43 (2005)
15. Hyperledger C Open Source Blockchain Technologies. https://www.hyperledger.org/
16. Kiffer, L., Rajaraman, R., Shelat, A.: A better method to analyze blockchain consistency. In: Proceedings of ACM Conference on Computer and Communications Security (CCS), pp. 729–744 (2018)
17. Lei, A., Cruickshank, H.S., Cao, Y., Asuquo, P.M., Ogah, C.P.A., Sun, Z.: Blockchain-based dynamic key management for heterogeneous intelligent transportation systems. IEEE Internet Things J. **4**(6), 1832–1843 (2017)
18. Li, Z., Chen, Y., Beach, A.: Towards scalable and robust distributed intrusion alert fusion with good load balancing. In: Proceedings of the 2006 SIGCOMM Workshop on Large-Scale Attack Defense (LSAD), pp. 115–122 (2006)

19. Li, W., Meng, Y., Kwok, L.-F.: Enhancing trust evaluation using intrusion sensitivity in collaborative intrusion detection networks: feasibility and challenges. In: Proceedings of the 9th International Conference on Computational Intelligence and Security (CIS), pp. 518–522. IEEE (2013)

20. Li, W., Meng, W., Kwok, L.-F.: Design of intrusion sensitivity-based trust management model for collaborative intrusion detection networks. In: Zhou, J., Gal-Oz, N., Zhang, J., Gudes, E. (eds.) IFIPTM 2014. IAICT, vol. 430, pp. 61–76. Springer, Heidelberg (2014). https://doi.org/10.1007/978-3-662-43813-8_5

21. Li, W., Meng, W.: Enhancing collaborative intrusion detection networks using intrusion sensitivity in detecting pollution attacks. Inf. Comput. Secur. **24**(3), 265–276 (2016)

22. Li, L., et al.: CreditCoin: a privacy-preserving blockchain-based incentive announcement network for communications of smart vehicles. IEEE Trans. Intell. Transp. Syst. **19**(7), 2204–2220 (2018)

23. Li, W., Meng, W., Kwok, L.-F., Ip, H.H.S.: PMFA: toward passive message fingerprint attacks on challenge-based collaborative intrusion detection networks. In: Chen, J., Piuri, V., Su, C., Yung, M. (eds.) NSS 2016. LNCS, vol. 9955, pp. 433–449. Springer, Cham (2016). https://doi.org/10.1007/978-3-319-46298-1_28

24. Li, W., Meng, W., Kwok, L.-F.: SOOA: exploring special on-off attacks on challenge-based collaborative intrusion detection networks. In: Au, M.H.A., Castiglione, A., Choo, K.-K.R., Palmieri, F., Li, K.-C. (eds.) GPC 2017. LNCS, vol. 10232, pp. 402–415. Springer, Cham (2017). https://doi.org/10.1007/978-3-319-57186-7_30

25. Li, W., Meng, W., Kwok, L.-F.: Investigating the influence of special on-off attacks on challenge-based collaborative intrusion detection networks. Future Internet **10**(1), 1–16 (2018)

26. Li, W., Tug, S., Meng, W., Wang, Y.: Designing collaborative blockchained signature-based intrusion detection in IoT environments. Future Gener. Comput. Syst. **96**, 481–489 (2019)

27. Li, W., Kwok, L.-F.: Challenge-based collaborative intrusion detection networks under passive message fingerprint attack: a further analysis. J. Inf. Secur. Appl. **47**, 1–7 (2019)

28. Makhdoom, I., Abolhasan, M., Abbas, H., Ni, W.: Blockchain's adoption in IoT: the challenges, and a way forward. J. Netw. Comput. Appl. **125**, 251–279 (2019)

29. Marr, B.: 5 Blockchain Trends Everyone Should Know About. https://www.forbes.com/sites/bernardmarr/2019/01/28/5-blockchain-trends-everyone-should-know-about/#30c1ab523bb9. Accessed 10 Apr 2019

30. Meng, Y., Kwok, L.F.: Enhancing false alarm reduction using voted ensemble selection in intrusion detection. Int. J. Comput. Intell. Syst. **6**(4), 626–638 (2013)

31. Meng, Y., Li, W., Kwok, L.F.: Towards adaptive character frequency-based exclusive signature matching scheme and its applications in distributed intrusion detection. Comput. Netw. **57**(17), 3630–3640 (2013)

32. Meng, W., Li, W., Kwok, L.-F.: An evaluation of single character frequency-based exclusive signature matching in distinct IDS environments. In: Chow, S.S.M., Camenisch, J., Hui, L.C.K., Yiu, S.M. (eds.) ISC 2014. LNCS, vol. 8783, pp. 465–476. Springer, Cham (2014). https://doi.org/10.1007/978-3-319-13257-0_29

33. Meng, W., Li, W., Kwok, L.-F.: EFM: enhancing the performance of signature-based network intrusion detection systems using enhanced filter mechanism. Comput. Secur. **43**, 189–204 (2014)

34. Meng, W., Li, W., Kwok, L.-F.: Design of intelligent KNN-based alarm filter using knowledge-based alert verification in intrusion detection. Secur. Commun. Netw. 8(18), 3883–3895 (2015)
35. Meng, W., Luo, X., Li, W., Li, Y.: Design and evaluation of advanced collusion attacks on collaborative intrusion detection networks in practice. In: Proceedings of the 15th IEEE International Conference on Trust, Security and Privacy in Computing and Communications (TrustCom 2016), pp. 1061–1068 (2016)
36. Meng, W., Li, W., Xiang, Y., Choo, K.K.R.: A Bayesian inference-based detection mechanism to defend medical smartphone networks against insider attacks. J. Netw. Comput. Appl. 78, 162–169 (2017)
37. Meng, W., Li, W., Kwok, L.-F.: Towards effective trust-based packet filtering in collaborative network environments. IEEE Trans. Netw. Serv. Manag. 14(1), 233–245 (2017)
38. Meng, W., Tischhauser, E.W., Wang, Q., Wang, Y., Han, J.: When intrusion detection meets blockchain technology: a review. IEEE Access 6(1), 10179–10188 (2018)
39. Nakamoto, S.: Bitcoin: a peer-to-peer electronic cash system (2008). http://bitcoin.org/bitcoin.pdf
40. Orcutt, M.: How secure is blockchain really? https://www.technologyreview.com/s/610836/how-secure-is-blockchain-really/. Accessed 22 Mar 2019
41. Papadopoulos, C., Lindell, R., Mehringer, J., Hussain, A., Govindan, R.: COSSACK: coordinated suppression of simultaneous attacks. In: Proceedings of the 2003 DARPA Information Survivability Conference and Exposition (DISCEX), pp. 94–96 (2003)
42. Pass, R., Shi, E.: Thunderella: blockchains with optimistic instant confirmation. In: Nielsen, J.B., Rijmen, V. (eds.) EUROCRYPT 2018. LNCS, vol. 10821, pp. 3–33. Springer, Cham (2018). https://doi.org/10.1007/978-3-319-78375-8_1
43. Porras, P.A., Neumann, P.G.: EMERALD: event monitoring enabling responses to anomalous live disturbances. In: Proceedings of the 20th National Information Systems Security Conference, pp. 353–365 (1997)
44. Roesch, M.: Snort: lightweight intrusion detection for networks. In: Proceedings of USENIX Lisa Conference, pp. 229–238 (1999)
45. Scarfone, K., Mell, P.: Guide to intrusion detection and prevention systems (IDPS). NIST Special Publication 800-94 (2007)
46. Snapp, S.R., et al.: DIDS (distributed intrusion detection system) - motivation, architecture, and an early prototype. In: Proceedings of the 14th National Computer Security Conference, pp. 167–176 (1991)
47. Sharma, V.: An energy-efficient transaction model for the blockchain-enabled internet of vehicles (IoV). IEEE Commun. Lett. 23(2), 246–249 (2019)
48. Singh, S., Ra, I.H., Meng, W., Kaur, M., Cho, G.H.: SH-BlockCC: a secure and efficient IoT smart home architecture based on cloud computing and blockchain technology. Int. J. Distrib. Sens. Netw. (in press). SAGE
49. Snort: An an open source network intrusion prevention and detection system (IDS/IPS). http://www.snort.org/
50. Steichen, M., Hommes, S., State, R.: ChainGuard - a firewall for blockchain applications using SDN with OpenFlow. In: Proceedings of International Conference on Principles, Systems and Applications of IP Telecommunications (IPTComm), pp. 1–8 (2017)
51. Symantec 2019 Internet Security Threat Report. https://www.symantec.com/security-center/threat-report. Accessed 22 Mar 2019

52. Tug, S., Meng, W., Wang, Y.: CBSigIDS: towards collaborative blockchained signature-based intrusion detection. In: Proceedings of The 1st IEEE International Conference on Blockchain (Blockchain) (2018)
53. Tuan, T.A.: A game-theoretic analysis of trust management in P2P systems. In: Proceedings of ICCE, pp. 130–134 (2006)
54. Vasilomanolakis, E., Karuppayah, S., Muhlhauser, M., Fischer, M.: Taxonomy and survey of collaborative intrusion detection. ACM Comput. Surv. **47**(4), 55:1–55:33 (2015)
55. Vigna, G., Kemmerer, R.A.: NetSTAT: a network-based intrusion detection approach. In: Proceedings of Annual Computer Security Applications Conference (ACSAC), pp. 25–34 (1998)
56. Wan, C., et al.: Goshawk: a novel efficient, robust and flexible blockchain protocol. In: Guo, F., Huang, X., Yung, M. (eds.) Inscrypt 2018. LNCS, vol. 11449, pp. 49–69. Springer, Cham (2019). https://doi.org/10.1007/978-3-030-14234-6_3
57. Wang, Y., Meng, W., Li, W., Liu, Z., Liu, Y., Xue, H.: Adaptive machine learning-based alarm reduction via edge computing for distributed intrusion detection systems. Concurr. Comput. Pract. Exp. (2019). https://doi.org/10.1002/cpe.5101
58. Wood, G.: Ethereum: a secure decentralised generalised transaction ledger. EIP-150 Revision (2016)
59. Wu, Y.-S., Foo, B., Mei, Y., Bagchi, S.: Collaborative intrusion detection system (CIDS): a framework for accurate and efficient IDS. In: Proceedings of the 2003 Annual Computer Security Applications Conference (ACSAC), pp. 234–244 (2003)
60. Wüst, K., Gervais, A.: Do you need a blockchain? In: CVCBT, pp. 45–54 (2018)
61. Yegneswaran, V., Barford, P., Jha, S.: Global Intrusion Detection in the DOMINO Overlay System. In: Proceedings of the 2004 Network and Distributed System Security Symposium (NDSS), pp. 1–17 (2004)
62. The Zeek Network Security Monitor. https://www.zeek.org/

AIoTS - Artificial Intelligence and Industrial Internet-of-Things Security

Enhancement to the Privacy-Aware Authentication for Wi-Fi Based Indoor Positioning Systems

Jhonattan J. Barriga A.[1,2] ⓘ, Sang Guun Yoo[1,2(✉)] ⓘ,
and Juan Carlos Polo[3]

[1] Facultad de Ingeniería de Sistemas, Escuela Politécnica Nacional,
Quito, Ecuador
{jhonattan.barriga, sang.yoo}@epn.edu.ec
[2] Smart Lab, Escuela Politécnica Nacional, Quito, Ecuador
[3] Departamento de Ciencias de la Computación, Universidad de las Fuerzas
Armadas ESPE, Sangolqui, Ecuador
jcpolo@espe.edu.ec

Abstract. Indoor location-based application and services based on Wi-Fi have serious problems in terms of privacy since attackers could track users by capturing their MAC addresses. Although several initiatives have been proposed by scientific community to properly address authentication by strongly preserving privacy, there are still improvements and steps that need to be developed as it is not clearly stated what would occur if a device is lost, stole or compromised. It has not been said how an affected user should proceed in such case. In this situation, this work provides an enhancement to a previous solution based on pseudo-certificates issued by third-party authorities for anonymous authentication of mobile devices. The proposed scheme provides privacy to users willing to remove a device that has been stolen or lost. The proposed system offers security while maintaining minimal cryptographic overhead.

Keywords: Privacy · Anonymous de-authentication ·
Indoor positioning system · IPS · WLAN

1 Introduction

Scientific community and industry show a great interest on improving the accuracy of Indoor Positioning Systems (IPS) [1–3] because it is an alternative to GPS and it could be applied to different areas of Internet of Things such as healthcare and social life [2, 4]. Even though several technologies could be applied to acquire indoor positioning data, Wi-Fi is the most used technology since it is widely used among private and public organizations [3, 5].

Different improvements of indoor positioning systems have been presented in different works; however, in most of cases, the privacy issue has been left aside [6, 7]. Similarly, a novel privacy-aware authentication system for Wi-Fi IPS was proposed in [8]. Likewise, there are other approaches looking for protecting the privacy of users based on lightweight solutions [9], Secure Two Party Communication (STPC) [10],

© Springer Nature Switzerland AG 2019
J. Zhou et al. (Eds.): ACNS 2019 Workshops, LNCS 11605, pp. 143–155, 2019.
https://doi.org/10.1007/978-3-030-29729-9_8

Physical Signatures [11] or application to change MAC address randomly [12]. Additionally, there are other solutions that use continuous authentication with the exclusion of identifiers from message headers [13] or propose an scheme based on a secret, token and biometrics [14]. As shown, many solutions have been proposed to perform authentication by preserving anonymity. However, the aforementioned related works, in spite of being secure and privacy conservatives, do not consider the whole scenario which is associated when a device has been lost, stolen or compromised and needs to be removed from the system. The user must have the power to perform such action as he/she is the owner of the device. Moreover, the intervention of a system administrator would cause more problems since the administrator could unsubscribe a valid user by mistake or since the user will have to follow an administrative procedure to request the device removal making the user to wait longer than expected.

The use of pseudo-certificates guarantees privacy because they do not store users' information and they have a validity time to prevent being reused after a certain period of time [8]. However, such security mechanism is not enough since if a malicious user manages to obtain a valid (loss or stolen) device, he/she will be able to access to the system as the device is a registered equipment with a valid set of credentials. For this reason, this paper intends to make an improvement of [8] by providing a simple but secure mechanism for users with valid credentials that allows users removing their devices from the system. To reach this goal, it is necessary to design a proper protocol with features that allow to remove every record of the device from the whole system without disclosing information and preserving privacy and anonymity. In summary, the major contribution of this paper is to enhance a protocol designed in a previous work [8] by proposing a novel initiative to securely remove a device of a particular individual that was previously registered in the system [3, 15].

The rest of the paper is structured as follows, Sect. 2 comprises a brief revision on authentication protocols for indoor positioning systems based on WLAN. Then, the proposed solution is explained with details in Sect. 3. Later, Sect. 4 analyzes the proposed solution in terms of security and performance. Finally, the paper is concluded in Sect. 5.

2   State of the Art

Scientific community has focused on addressing privacy issues by proposing different works [8–17]. Reference [5] discusses several of those proposals including solutions based on obfuscation of sensitive data and usage of random MAC addresses. In [13] a proximity based control is proposed where the authors emphasize on removing previously generated information over packets to preserve privacy. An approach focused on a triple combination of a token, secret and biometrics is discussed in [14]. Such work is mainly oriented to use a smart phone to authenticate and authorize over a location-based system; in such system, two protocols are proposed: one for registration and another to handle authorization and authentication. Although the solution addresses privacy and authentication in a secure way, it does not describe how to remove a device that has been lost or stolen i.e. compromised. Likewise, the reference [10] indicates that the IPS server could violates user's privacy and that the device could forces the system to disclose its location. Their solution suggests the use of a Secure Two-Party Computation (STP) to protect the privacy of all the involved participants. In this approach, the user encrypts and sends the private inputs (RSSI distance measured from APs) to the server with a secure algorithm. However, this approach does not deliver an authentication mechanism before attaching to the system which means that any user would be able to send his/her inputs. Additionally, in case that a user wanted to be removed from the system, there is no formal process to achieve such goal. Indeed, lack of authentication might lead that the user would send its position to a rogue IPS server that might be deployed within the same network.

Moreover, the proposal discussed in [16] presents an algorithm called *TemporalVectorMap* (TVM). It allows a user to accurately know his/her position by taking advantage of a k-Anonymity Bloom (*kAB*) filter and a *bestNeighbors* generator of camouflaged localization requests. According to the authors, both of the aforementioned techniques seem to be resilient to some privacy attacks. This proposal draws two phases: (i) initial localization and (ii) continuous localization. During the first phase, the *kAB* is built based on the MAC address of an Access Point (AP), assuming that the AP is valid within the context of the network as it has been registered by the server. Phase of continuous location implements *bestNeighbors* algorithms to handle users that could be moving around the deployed ecosystem. Authors suggests that their solution is not prone to linking attack as there are no attribute records stored in the server and it is resistant to homogeneity attack as it uses hashing to generate a set of unique AP MAC values. Anyhow, along this solution, there is no formal procedure for registering or removing a device.

Another approach based on WLAN for wireless sensor networks is discussed in [9]. The authors propose a lightweight authentication protocol which is mainly based on Fermat Number Transform (FNT) and Chinese Remainder Theorem (CRT) to maintain secure communication. The encryption and decryption algorithm are based on a protocol that involves a combination of substitution cipher and columnar transposition which withstands linear crypt analysis rather than using a formal known one. The authentication relies on generating a prime number that is stored in the node and in the server. The node sends an authentication request to the server which is processed and validated at the base station. Although this schema considers a secure authentication schema, it does not deliver an optimal process for compromised node removal since the administrator has to remove the compromised node's key from the base station.

Furthermore, the solution named as "IMAKA-Tate" [17] proposes an schema based on three-way handshake mutual authentication and key agreement in conjunction with authentication against an Extensible Authentication Protocol (EAP). Mutual Authentication and Key Agreement are used so that each participant generates random challenge, which is encrypted by the corresponding public key of the recipient. This work properly addresses anonymity and privacy, but it does not include a formal procedure for device removal.

The solution discussed in [11] proposes an authentication protocol IPS based in WLAN that verifies user information based on the physical layer (PHY) signatures within WLAN preambles. It mainly uses Carrier Frequency Offset (CFO) and multipath plus Channel State Information (CSI) to protect wireless communications since the handshake phase between the mobile users and the access point (AP), and whilst validating the truthfulness of a reported location from a user of the system. In this current solution, there is no need of credentials for registering the user as everything is handled at the PHY layer. This solution, like the previous ones, lacks from having a formal procedure to remove an undesired device from the system (remove from authentication system).

A privacy protection mechanism for indoor positioning is presented in [12]. This mechanism proposes the use of an application that changes the MAC address of the phone periodically. They use this approach to provide privacy to the user as the server will not determine his/her identity. Although privacy is protected, there might be a potential issue if a MAC address is repeated along two users handling the same manufacturer phone. The process of authentication is not formally defined, but it appears that the application installed will be in charge on performing such action. Again, this solution, like the previous ones, does not deliver a formal process to remove the device from the system rather than uninstalling the application from the phone.

Weaknesses about PriWFL are exploited and discussed in [18]. These weaknesses might let attackers to obtain the position of a user. The authors present a practical Server Data Privacy Attack where they point that an attacker only needs to obtain a pair of distances. They also discuss an attack that reveals the order of RSS values. As stated by the authors there are non-trivial problems that may dramatically affect the localization accuracy. Furthermore, the authors propose Fully Homomorphic Encryption and Somewhat Homomorphic Encryption but they are computational costly or impractical for Wi-Fi schema. Secure Multiparty Computation (MPC) is analyzed but as reviewed by the authors it may generate communication overhead. Paillier PKE is analyzed from two perspectives (Signs of Differences and Garbled Circuits), where the first approach seems secure but might be susceptible to order attack, whilst the second approach is more secure as it resilient to Client Privacy Attacks (scenarios 1 and 2) as the attacker could not infer the location of a client if the secret key is not known. Likewise, if the MPC is secure and the randomness are fresh, an attacker cannot learn combined distances. The inclusion of Paillier encryption let a client to learn only signs of distances. The solution proposed clearly analyze and exploit weaknesses focusing on an attacker compromising the database of a provider. This paper makes good points on preserving privacy.

Practical Privacy-Preserving Indoor Localization using OuTsourcing (PILOT) is another approach which focuses like the previous work on preserving privacy in an Indoor Positioning System [19]. Semi Trusted Third Parties (STTPs), a client and an Indoor Location Provider (ILP) are involved in the approach described by the authors. In the described scenario the client collects signal strengths from access points predefined by the ILP and then shared to the STTPs by a secure channel. Every STTP calculates an ILP protocol by using a Secure Two-Party Computation (STPC). The solution proposed is secure against semi-honest non-colluding STTPs, malicious clients and ILP servers. According to the authors, the use of the ABY-Framework ensures that intermediate secret-shares are secure as well as conversions, and final target location. The proposed schema guarantees that if one STTP and the client are not compromised; then, the client will not be able to determine its location. Likewise, if the ILP and one STTP are not corrupted and no matter if there is a leak of information, it will not be possible to determine the RSSs of the database of the ILP server. In regards of connectivity this approach relies on secure communication. The main contribution of this paper is a protocol that deals with most of communication and computation on third-parties (STTPs) rather than the mobile client as it poses limited hardware resources. Although this contribution shows a strong on privacy-preserving schema, it does not present as a use case where a device has been compromised or stolen and the user would have the power to act in such case.

Another solution is discussed in [8], where the use of pseudo-certificates helps to provide privacy and anonymity to the user attached to the system. In this proposal, the user first has to register his device, and then the system will generate a set of pseudo-certificates with an expiry time. These certificates will let a server to determine a user position without knowing/revealing its identity. Although this approach describes the process of authentication, it does not handle the process of removing a device.

We have examined several solutions in regards of authentication procedures for IPS based on WLAN. All of them showed the need to have a formal and secure process for removing devices that have been lost or stolen. With this antecedent, our proposal is to perform and enhancement to [8], by adding a formal and secure process for removing devices, giving the user the right to perform such action without compromising his/her privacy. The proposed protocol will be described in detail and analyzed from a security and performance perspectives in the next sections.

3   Proposed Protocol

3.1   Overview of the System

Since the objective of this work is to deliver the mobile device removal process to a previous work, the proposed mechanism uses the same three main entities for Authentication, Authority and Accounting (AAA). A brief overview of the previous work that will be enhanced is shown below (see Fig. 1). The reviewed system is composed by three main entities:

(1) User environment composed of the user and his/her mobile device(s).
(2) A Certificate Authority (CA) which manages the accounts of users, data of their mobile devices, and devices' pseudo-certificates/private keys.
(3) An IPS Server that provides the indoor positioning service, which is registered in the CA.

For a better understanding of the system, we recommend to refer to the previous work [8].

3.2   Proposal of Device Removal Functionality: An Overview

In a previous work [8], a protocol for providing privacy by using an anonymous authentication schema was designed. However, it does not have a process to remove a previously registered mobile device, which means that a lost or stolen mobile device

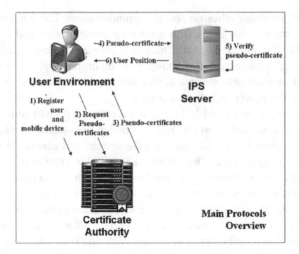

Fig. 1. Overview of the system

can be used by an illegal/malicious user. In this sense, this work enhances the previous work by adding the mobile device removal process which contains the following steps (see Fig. 2).

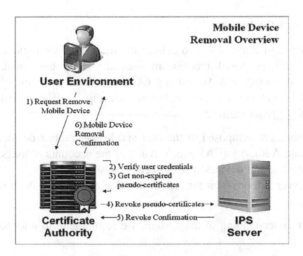

Fig. 2. Proposed enhancement schema

The user who wants to remove his/her device must be previously registered to the Certificate Authority (CA). The user first authenticates using his/her credentials. The CA validates credentials and if successful, it returns a list of available devices registered by such user. The user selects the devices to be removed and submits a request to remove the mobile device. Then, the CA will get all non-expired pseudo-certificates and will perform a request for revoking the pseudo-certificates from the Indoor Positioning System (IPS) Server. For this, the CA sends a list of certificates that need to be revoked by the IPS Server and the IPS server validates the request's authenticity, revokes the pseudo-certificates and confirms the revocation procedure to the CA. Then, the CA removes all the pseudo-certificates associated to the device that needs to be removed. Finally, the user receives a removal confirmation message.

3.3   Proposal of Device Removal Functionality: Details

In the previous subsection, we have described briefly the flow of the proposed system. Now, this subsection will describe the details of the proposed functionality. The notation used to describe the protocol is detailed in Table 1.

Table 1. Notations used in the proposed solution

Notation	Description
U_i	i^{th} user
MD_{j_Ui}	U_i's j^{th} mobile device
$RN1, RN2,..., RNn$	Random nonces
$RK1, RK2...RKn, RK_{CA}, RK_{IPS}$	Random symmetric keys
CA	Certificate Authority
$Pubkey_{CA}, Prikey_{CA}$	CA's asymmetric key pair
$Pubkey_{IPS}, Prikey_{IPS}$	$IPS\ Server$'s asymmetric key pair
ID_{Ui}	Identification of U_i
PW_{Ui}	Password of U_i
$NAME_{MDj\_Ui}$	Name of MD_{j_Ui}
MAC_{MDj_Ui}	MAC address of MD_{j_Ui}
$\{PCert_{(CA,MDj\_Ui)1},..., PCert_{(CA,MDj\_Ui)n}\}$	Pseudo-certificates of MD_{j_Ui}
$\{Prikey_{(CA,MDj\_Ui)1},..., Prikey_{(CA,MDj\_Ui)n}\}$	Private keys of pseudo-certificates of MD_{j_Ui}
IP_{IPS}	$IPS\ Server$'s IP address
$PCert_{(CA,MDj\_Ui)k}$	k^{th} (unused) pseudo-certificate
Pos_{MDj_Ui}	Current position of MD_{j_Ui}
$\|$	String concatenation
$h(.)$	One-way hash function
$AEnc(x, y)$	Asymmetric encryption of message y using the key x
$ADec(x, y)$	Asymmetric decryption of message y using the key x
$SEnc(x, y)$	Symmetric encryption of message y using the key x
$SDec(x, y)$	Symmetric decryption of message y using the key x
$Sign(x, y)$	Digital signature of message y using the private key x
$VerifySign(x, y)$	Digital signature verification of signature y using public key x

Mobile Device Removal. This protocol is executed as follows (see Fig. 3). First, the user U_i inputs his/her identity ID_{Ui} and password PW_{Ui} to his/her mobile device MD_{j_Ui}. Then, MD_{j_Ui} communicates with the third-party CA and asks for user authentication. After receiving the request message, CA generates a random number $RN8$ and sends it to MD_{j_Ui}. Once received the response from CA, MD_{i_Ui} generates a random nonce $RN9$, a random symmetric key $RK4$, and calculates $M12 = AEnc$ $(Pubkey_{CA}, RK4)$ and $M13 = SEnc(RK4, RN8||RN9||ID_{Ui}||h(PW_{Ui}))$, where $AEnc(x, y)$ is an asymmetric encryption of message y using the key x, $Pubkey_{CA}$ is CA's public key, $SEnc(x, y)$ is a symmetric encryption of message y using the key x, $||$ is a concatenation operation, and $h(.)$ is a one-way hash function. Once calculated $M12$ and $M13$, MD_{j_Ui} sends those values to CA.

On the other side, CA gets $RK4$ by executing $ADec(Prikey_{CA}, M12)$ where $ADec(x, y)$ is an asymmetric decryption of an encrypted message y using the key x, and uses $RK4$ to get $RN8'$, $RN9'$, ID_{Ui}, and $h(PW_{Ui})$ by executing $SDec(RK4, M13)$, where $SDec(x, y)$ is a symmetric decryption of an encrypted message y using the key x. Once gotten $RN8'$, CA verifies the freshness of the message by comparing the decrypted $RN8'$ with the random nonce created previously by itself i.e. $RN8$. This step allows CA to protect against replay attacks. After verifying the validity of the message, CA verifies if ID_{Ui} and $h(PW_{Ui})$ are valid credentials otherwise the process is aborted. If credentials are valid, the CA retrieves a list of registered devices from DB that belong to the user ID_{Ui}, this list is a collection of tuples formed by the $NAME$ of the device and its MAC $Address$ $\{(NAME_{MD1\_Ui}, MAC_{MDn\_Ui}),..., (NAME_{MDn\_Ui}, MAC_{MDn\_Ui})\}$. Then, the CA generates a random nonce $RN10$, and calculates $M14 = SEnc(RK4, RN9'||RN10||\{(NAME\text{-}_{MD1\_Ui}, MAC_{MDn\_Ui}),..., (NAME_{MD1\_Ui}, MAC_{MDn\_Ui})\})$, which is sent to the mobile device MD_{j_Ui}.

The mobile device MD_{j_Ui} gets $RN9''||RN10||\{(NAME_{MD1\_Ui}, MAC_{MDn\_Ui}),..., (NAME_{MD1\_Ui}, MAC_{MDn\_Ui})\}$ by excecuting $SDec(RK4, M14)$. Once gotten $RN9''$, the mobile device verifies the freshness of the message by comparing the decrypted $RN9''$ with the random nonce generated previously by itself i.e. $RN9$. After verifying the authenticity of the message, the mobile device generates $M15 = \{(NAME_{MD1\_Ui}, MAC_{MD1\_Ui}),..., (NAME_{MDn\_Ui}, MAC_{MDn\_Ui})\}$, and display the list of registered devices to the user.

The user U_i selects the registered device from the list ($M15$) that wants be removed (MAC_{MDg_Ui}). The mobile device generates a random nonce $RN11$, and calculates $M16 = SEnc(RK4, RN10'||RN11||MAC_{MDg\_Ui})$, and sends $M16$ to the CA.

The CA gets $RN10''||RN11||MAC_{MDg\_Ui}$ by executing $SDec(RK4, M16)$. Once gotten $RN10''$, the CA verifies the freshness of the message by comparing the decrypted $RN10''$ with the random nonce created before by itself i.e. $RN10$. If such values are the same, the CA gets all the not expired pseudo-certificates of the mobile device that are stored in the DB $\{PCert_{(CA,MDj\_Ui)1},..., PCert_{(CA,MDj\_Ui)n}\}$. Then, the CA submits a request to the IPS Server and it generates a random nonce $RN12$ which is sent back to the CA. The CA, generates and random nonce $RN13$, a random key RK_{CA} and calculates $M17$ and $M18$, where $M17 = AEnc(PubKey_{IPS}, RK_{CA})$. $M18 = SEnc$ $(RK_{CA}, RN12||RN13||\{PCert_{(CA,MDj\_Ui)1},..., PCert_{(CA,MDj\_Ui)n}\}$ $|| Sign(PriKey_{CA}, RK_{CA}))$, and $Sign(x, y)$ is the signing function of a message y using the private key x. Once calcultated $M17$ and $M18$, the CA sends those messages to the IPS $Server$.

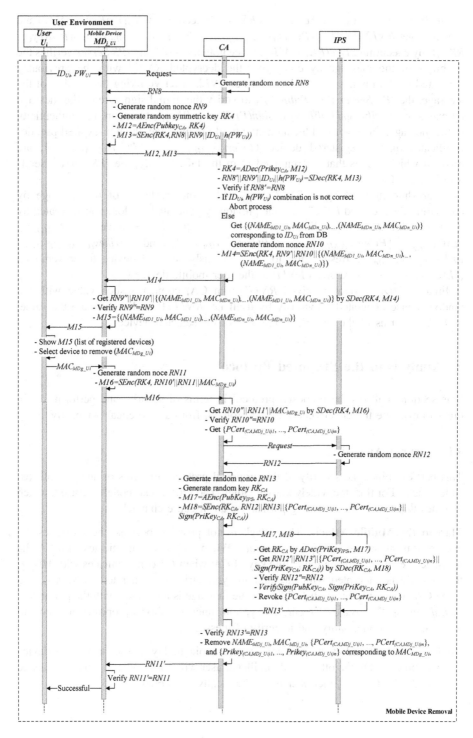

Fig. 3. Mobile device removal protocol

On the other side, the *IPS Server* gets RK_{CA} by executing $ADec(PriKey_{IPS}, M17)$ and uses it to get $RN12''\|RN13'\|\{PCert_{(CA,MDj\_Ui)1},\dots, PCert_{(CA,MDj\_Ui)n}\}\|Sign(PriKey_{CA}, RK_{CA}))$ by executing $SDec(RK_{CA}, M18)$. Once gotten $RN12''$, the *IPS Server* verifies the freshness of the message by comparing the decrypted $RN12''$ with the previously generated random nonce created by itself i.e. $RN12$. After verifying the validity of the message, the *IPS Server* uses $PubKey_{CA}$ to verify the digital signature of the message by executing $VerifySign(PubKey_{CA}, Sign(PriKey_{CA}, RK_{CA}))$ to ensure the authenticity of the message. Once verified the authenticity of the message, the non-expired pseudo-certificates of the registered device $\{PCert_{(CA,MDj\_Ui)1},\dots, PCert_{(CA,MDj\_Ui)n}\}$, are revoked which means that are removed from the DB. Finally, the *IPS Server*, sends $RN13'$ to the CA.

Meanwhile, the CA, gets $RN13'$ and verifies the freshness of the message by comparing the received $RN13'$ with the previously generated random nonce created by itself i.e. $RN13$. After validating the message, the CA removes $NAME_{MDj\_Ui}$, MAC_{MDj_Ui}, $\{PCert_{(CA,MDj\_Ui)1},\dots, PCert_{(CA,MDj\_Ui)n}\}$, and $\{Prikey_{(CA,MDj\_Ui)1},\dots, Prikey_{(CA,MDj\_Ui)n}\}$ corresponding to the previously selected device to be removed (MAC_{MDg_Ui}.). The CA, sents $RN11'$ to the user mobile device.

Finally, MD_{j_Ui}, once received $RN11'$ from CA, compares such value with the random nonce generated previously by itself i.e. $RN11$. If such values are the same, MD_{j_Ui}, confirms U_i the successful removal of the selected device.

4   Analysis of the Proposed Protocol

This section analyzes the proposed protocol in terms of security and performance in order to evaluate the effectiveness of the protocol from a theoretical perspective.

4.1   Security Analysis

This part examines the security of the proposed protocol in terms of analysis of possible attacks. For this, the widely known Dolev-Yao [20] threat model was used, which assumes that two communicating parties use an insecure channel.

Man in the Middle Attack. This attack is not possible because the messages are encrypted using secure encryption functions. When MD_{j_Ui} communicates with *CA*, the message is encrypted using the public key of *CA*; when *CA* communicates with MD_{j_Ui} the message is encrypted with the random symmetric key generated by MD_{j_U}; and when CA communicates with *IPS Server* the message is encrypted with the public key of $PubKey_{IPS}$. The usage of secure encryption functions allows proposed protocols to maintain the confidentiality and integrity of messages.

Replay Attack. Random nonces are used to avoid replay attacks in mobile device removal process. On the other hand, an illegal user will not be able to remove a device because the MAC_{MDg_Ui} identifier is symmetrically encrypted.

Password Guessing Attack. PW_{Ui} is not stored anywhere. Instead, a variant value $h(PW_{Ui})$ is used for user validation. Since $h(.)$ is a secure one-way hash function, the attacker cannot guess the PW_{Ui} from $h(PW_{Ui})$. Hence, this attack is not possible.

Privileged-Insider Attack. In the proposed solution, MD_{j_Ui} never transmits the password of the user PW_{Ui} in plaintext. Instead, a variant value $h(PW_{Ui})$ is sent to the CA. Even a privileged-insider of CA cannot guess the PW_{Ui} because $h(PW_{Ui})$ is calculated using a secure one-way hash function. Also, a malicious user might try to revoke certificates from a valid device; however, such malicious user will have to manually generate requests to the *IPS Server* to obtain approval for a complete removal.

Brute Force Attack. The attacker can attempt to remove a valid device by sending random or sequential messages to the CA. However, the use of random nonces helps to prevent this attack.

Separation of Responsibilities. *CA* manages only the information of the users/mobile devices while *IPS Server* manages only the information about the relation between a pseudo-certificate and position.

4.2   Performance Analysis

Table 2 indicates the overhead of cryptographic steps of the proposed protocol. It is important to mention that the cryptographic overhead in each protocol is minimal; therefore, it does not affect to the real implementation of the proposed solution.

Table 2. Cryptographic overhead (i.e. number of operations)

Phase	Entity	Proposed
Mobile Device Removal	MD_{j_Ui}	1 AEnc + 2 SEnc + 1 H + 1 SDec
	CA	2 SEnc + 1 ADec + 1 SDec + 1 AEnc + 1 Sign
	IPS Server	1 ADec + 1 SDec + 1 VerifySign

AEnc: Asymmetric encryption, *ADec*: Asymmetric decryption, *H*: hash, *SEnc*: Symmetric encryption, *SDec*: Symmetric decryption, *Sign*: Creation of digital signature, *VerifySign*: Verification of digital signature.

5   Conclusions and Future Direction

This paper has proposed an enhancement to the novel authentication system for Indoor Positioning Systems that includes a protocol for removing mobile devices. The proposed solution allows a user to remove his/her registered devices if they have been stolen or lost, so that an illegal user will not be able to use it. The proposed solution still provides a secure authentication system for IPS while maintaining a minimal performance overhead. The proposed approach gives to the user the power to securely remove his/her device without the intervention of a third-party, reducing the risk of involuntary mistakes. In the near future, we will continue our research by implementing the suggested protocol in a real scenario and extending to more IoT devices.

Acknowledgements. The authors gratefully acknowledge the financial support provided by the Escuela Politécnica Nacional, for the development of the project PIJ-17-08 – "Diseño e Implementación de un Sistema de Parqueadero Inteligente".

References

1. Farid, Z., Nordin, R., Ismail, M.: Recent advances in wireless indoor localization techniques and system. J. Comput. Netw. Commun. **2013** (2013)
2. He, S., Chan, S.H.G.: Wi-Fi fingerprint-based indoor positioning: Recent advances and comparisons. IEEE Commun. Surv. Tutorials **18**(1), 466–490 (2016)
3. Stojanović, D.H., Stojanović, N.M.: Indoor localization and tracking: Methods, technologies and research challenges. Facta Univ. Ser.: Autom. Control Robot. **13**(1), 57–72 (2014)
4. Deng, Z., Yu, Y., Yuan, X., Wan, N., Yang, L.: Situation and development tendency of indoor positioning. China Commun. **10**(3), 42–55 (2013)
5. Lohan, E.: User privacy risks and protection in WLAN-Based (2015)
6. Mendez, D., Papapanagiotou, I., Yang, B.: Internet of things: survey on security and privacy. Inf. Secur. J. Glob. Perspect. 1–16 (2017)
7. Chen, L., et al.: Robustness, security and privacy in location-based services for future IoT: a survey. IEEE Access **5**, 8956–8977 (2017)
8. Yoo, S.G., Barriga, J.J.: Privacy-aware authentication for Wi-Fi based indoor positioning systems. In: Batten, L., Kim, D.S., Zhang, X., Li, G. (eds.) ATIS 2017. CCIS, vol. 719, pp. 201–213. Springer, Singapore (2017). https://doi.org/10.1007/978-981-10-5421-1_17
9. Shah, M.D., Gala, S.N.: Lightweight authentication protocol used in wireless sensor network. In: 2014 International Conference on Circuits, Systems, Communication and Information Technology Applications (CSCITA) Lightweight, pp. 138–143 (2014)
10. Ziegeldorf, J.H., Viol, N., Henze, M., Wehrle, K.: POSTER : privacy-preserving indoor localization. In: 7th ACM Conference on Security and Privacy in Wireless and Mobile Networks (WiSec 2014), Oxford, United Kingdom (2014)
11. Wang, W., Chen, Y., Zhang, Q.: Privacy-preserving location authentication in Wi-Fi networks using fine-grained physical layer signatures. IEEE Trans. Wirel. Commun. **15**, 1218–1225 (2015)
12. Kim, S., Yoo, S.G., Kim, J.: Privacy protection mechanism for indoor positioning systems. Int. J. Appl. Eng. Res. **12**(9), 1982–1986 (2017)
13. Agudo, I., Rios, R., Lopez, J.: A privacy-aware continuous authentication scheme for proximity-based access control. Comput. Secur. **39**, 117–126 (2013)
14. Zhang, F., Kondoro, A., Muftic, S.: Location-based authentication and authorization using smart phones. In: 2012 IEEE 11th International Conference on Trust, Security and Privacy in Computing and Communications, pp. 1285–1292 (2012)
15. Fawaz, K., Kim, K., Shin, K.G.: Privacy vs. reward in indoor location-based services. Proc. Priv. Enhancing Technol. **2016**(4), 102–122 (2016)
16. Konstantinidis, A., Chatzimilioudis, G., Zeinalipour-Yazti, D., Mpeis, P., Pelekis, N., Theodoridis, Y.: Privacy-preserving indoor localization on smartphones. IEEE Trans. Knowl. Data Eng. **27**(11), 3042–3055 (2015)
17. Fal Sadikin, M., Kyas, M.: IMAKA-Tate: secure and efficient privacy preserving for indoor positioning applications, vol. 5760, March 2016

18. Yang, Z., Järvinen, K.: The death and rebirth of privacy-preserving WiFi fingerprint localization with paillier encryption (full version). In: IEEE International Conference on Computer Communications 2018 (INFOCOM 2018), p. 21 (2018)
19. Järvinen, K., et al.: PILOT: practical privacy-preserving indoor localization using outsourcing. In: 4th IEEE European Symposium on Security and Privacy, p. 16 (2019)
20. Dolev, D., Yao, A.: On the security of public key protocols. IEEE Trans. Inf. Theory **29**(2), 198–208 (1983)

Design of a FDIA Resilient Protection Scheme for Power Networks by Securing Minimal Sensor Set

Tanmoy Kanti Das[1][(✉)], Subhojit Ghosh[2], Ebha Koley[2], and Jianying Zhou[3]

[1] Department of Computer Applications, National Institute of Technology Raipur,
G.E. Road, Raipur 492010, Chhattisgarh, India
`tkdas.mca@nitrr.ac.in`
[2] Department of Electrical Engineering, National Institute of Technology Raipur,
G.E. Road, Raipur 492010, Chhattisgarh, India
`{sghosh.ele,ekoley.ele}@nitrr.ac.in`
[3] Singapore University of Technology and Design, Singapore 487372, Singapore
`jianying_zhou@sutd.edu.sg`

Abstract. Recent times have witnessed increasing utilization of wide area measurements to design the transmission line protection schemes as wide area measurements improve the reliability of protection methods. Usage of ICT tools for communicating sensor measurement in power networks demands immunity and resiliency of the associated protection scheme against false data injection attack (FDIA). Immunity against malicious manipulation of sensor information is attainable by securing the communication channels connecting the sensors through cryptographic protocols, and encryption. However, securing all the sensors and communication channels is economically unviable. A practical solution involves securing a reduced set of sensors without compromising fault detection accuracy. With the aim of developing a simple, economically viable and FDIA resilient scheme under the assumption that the adversary has complete knowledge of the system dynamics, the present work proposes a logical analysis of data (LAD) based fault detection scheme. The proposed scheme identifies the minimal set of sensors for FDIA resiliency and detects the state (faulty or healthy) of the power network relying on the measurements received from the 'minimal sensor set' only. Validation of the proposed protection scheme on IEEE 9-bus system reveals that in addition to being FDIA resilient, it is reliable and computationally efficient.

Keywords: Smart grid · Transmission line protection · False Data Injection Attack (FDIA) · Fault detection · Partially defined Boolean function (pdBf) · Logical analysis of data

1 Introduction

The reliable operation of any power system is heavily dependent on the development of a suitable protection scheme against line faults and contingencies.

© Springer Nature Switzerland AG 2019
J. Zhou et al. (Eds.): ACNS 2019 Workshops, LNCS 11605, pp. 156–171, 2019.
https://doi.org/10.1007/978-3-030-29729-9_9

A reliable protection scheme allows for faster fault detection and hence early restoration of power supply post-fault. In recent times, with the soaring assimilation of the physical power transmission system with the cyber information and communication tools in smart grids, the possibility of cyber-attacks poses a serious challenge towards the development and implementation of a reliable protection mechanism against faults. The protection component plays a significant role in the overall operation and control of a power system. The increased stress on rapid detection of faults and reduction in fault levels is arising because of the penetration of renewable energy sources has led to a paradigm shift from classical protection scheme using local measurements to protection scheme relying on 'wide area measurements' [1]. The effective performance of a protection scheme, which rely on wide area measurements, is highly dependent on the sensor information transmitted to the control centers through the cyber network. Over-dependence of power systems on the public communication networks for reliable monitoring and operation, makes it vulnerable to cyber attacks [2].

False data injection attack (FDIA) is considered as the most potent cyber attack in which the overall power grid can be made to collapse by the hacker with minimal effort. During FDIA, the attacker corrupts the integrity of a set of measurements that are used in the protection algorithm by tampering the meter/sensor measurements [3,4]. The protection algorithms are part of the backup protection strategy, which is operated from the control center(s). Transmission of false data to the control center may lead to unnecessary control action that might result in contingencies or even blackout. Consequently, the present scenario demands a protection scheme that is either immune to data falsification or/and includes a component for preemptive detection of false data injection. The state-of-the-art for protection of transmission lines [5–8] has not addressed the deployment of a security mechanism against vulnerabilities caused by FDIA.

Conventional power networks address the need for system monitoring through state estimation [9], which is carried out using the power system model, and sensor informatics. Conventional bad data detection methods that are part of state estimators are supposed to detect any malicious manipulation of sensor information. However, Liu et al. [10] have demonstrated that a hacker, having enough knowledge about the system dynamics, can bypass the bad data detection techniques and inject any arbitrary errors into state variables by suitably injecting malicious sensor information using FDIA. Thus, the manipulation of sensor information during an attack can provide a deceptive picture regarding the system dynamics and operation, leading to either non-operation of the relay during fault or tripping of the relay followed by isolation during a non-faulty/healthy case. Inappropriate actions of the protective relays, and a delay in the detection of such attacks can result in a huge economic loss, asset damage, and collapse of the related sub-systems and control mechanisms. With the explosive growth in the use of sensors (CT, PT, PMU) and communication network for continuous online real-time monitoring using the information of the current and/or voltage signals at different buses or locations, the scope of mounting a false data injection attack has increased significantly in recent times.

Recent works on FDIA in power grid have mainly concentrated on the modeling of FDIA, detection of an attack and defensive measures [11–25]. The probable implications arising out of FDIA on power system have been addressed in [3,11,12]. The notable schemes reported for FDIA detection in power networks are based on transmission line susceptance measurements [13], reactance perturbation [14], joint-transformation [15], extreme learning machine [16], sparse optimization [17] and cumulative sum approach [18]. Yang et al. [19] proposed a countermeasure to FDIA using the premise that the *sensors*, which measure injective power flow in the buses and are connected to several other buses require security. Since inaccessibility of those *sensors* will make it difficult for an attacker to mount an FDIA. A defense mechanism to protect a set of state variables has been proposed in [20,21].

An adaptive Markov based defense strategy for the protection of smart grid has been reported in [22]. A two-layer attack-defense mechanism to protect PMUs against FDIA is presented in [23]. In [24], a greedy search algorithm is presented to obtain the subset of measurements required to be protected to defend against FDIA. In [25], a scheme based on bilevel mixed integer linear programming has been presented to prevent the falsification of load data. A defensive method against data integrity attacks based on the optimal PMU placement strategy is proposed in [23]. An algorithm for appropriate placement of PMU in electric transmission network for reliable state estimation against FDIA has been presented in [26]. In [27], a generalized scheme for detecting data integrity attacks in cyber-physical systems based on sensor characteristics and noise dynamics has been proposed. Stackelberg game has been used for detector tunning to detect FDIA in [28].

Most of the existing works on FDIA mentioned above have only concentrated on detection of FDIA without analyzing its effect on the operation of the transmission line protection module. To the best of our knowledge, no work has been reported on analyzing the implications of FDIA on fault detection and developing an FDIA immune protection scheme. A simple solution to this problem is to replace the set of all the existing sensors τ with 'secure sensors'. Secure sensors communicate using cryptographic protocols and methods, which prevent any chances of FDIA unless the keys of cryptographic protocols are compromised. Moreover, secure sensors are protected from physical tampering using tamper-resistant hardware. However, the sheer number of installed meters/sensors in electric grids makes it impractical to replace all the sensors with secured sensors [29]. At best, we can secure a small set of sensors τ_s, where $\tau_s \subset \tau$.

Moreover, the selection of the reduced sensor set should ensure no degradation in the performance of the protection algorithm, in terms of accurately detecting various fault scenarios, even in the presence of FDIA. This demands optimally locating those sensors whose information either do not contribute to the system monitoring or can be correlated with other sensor information. With τ_s, the protection algorithm is expected to carry out the intended task of detecting the faults by suitable mapping of secured sensor information with the fault scenarios. Considering a moderate-size network having a few hundred installed sensors, a brute force search to locate τ_s over all possible small-size subsets are prohibitively costly.

In the present work, the twin problems of identifying τ_s and correlating the protection scheme output (faulty/healthy) with the sensor information are solved using a classifier, which utilizes a partially defined Boolean function based data analysis technique known as Logical Analysis of Data (LAD) [30–32]. For a two-class classification problem, LAD aims at optimally generating a set of rules/patterns, which can collectively classify all the known observations (power system scenarios). Features/sensor-measurements, which contribute insignificantly to the classification task, are ignored and further not included in the rules. In addition of providing immunity against FDIA, a significant contribution of the LAD based protection scheme is the reduction in complexity of the detection algorithm since the overall sensor information is substantially reduced without employing any dimension reduction technique.

Popular classifiers, like KNN, ANN, SVM, etc., which are generally preceded by some feature extraction method, are difficult to implement on the digital relays that work on threshold settings. On the contrary, in the LAD-based scheme, the raw data (i.e., sensor information) are directly fed into the classifier without any pre-processing and the classification rules provide a threshold for each input feature (i.e., sensor information in the present problem). Further, the generalization of LAD to datasets of varying dimensions makes the proposed scheme independent of the power system network topology. It is to be noted that, unlike the existing works on 'optimal sensor placement' [23] based on maintaining *'system state observability'*, the present work aims at the identification of optimal sensor set for imparting immunity to protection scheme against FDIA. Securing sensors identified using 'system state observability' do not guarantee immunity to power line protection schemes against FDIA.

The effectiveness of the proposed scheme has been evaluated by performing extensive simulations under normal operation and FDIA for IEEE 9 bus system. While simulating the false data injection attack, it is assumed that the attacker has complete knowledge regarding the power system model. For varying scenarios, the proposed scheme is able to correctly detect the state of transmission line, i.e., faulty or healthy under FDIA of varying degrees with significantly reduced execution time (maximum $45\,\mu s$). The highlights/novelty of the proposed work can be summarized as:

1. Development of an FDIA immune protection scheme with the assumption that the *attacker has complete knowledge of the power system.*
2. Development of a data analysis based approach for identifying the limited set of sensors that would be secured using tamper-resistant hardware, cryptographic protocol, and encryption.
3. Design of a rule-based fault detection scheme by mapping the secured sensor information with the state of the power system using LAD-based classifier.

The remainder of the paper is organized as follows. Section 2 discusses the development of the proposed LAD based protection scheme. Section 3 demonstrates the test results on the IEEE 9-bus system to exemplify the proposed scheme. Finally, Sect. 4 summarizes the contributions of the paper and provides conclusions and future research direction.

2   Design of a LAD Based Classifier for Fault Detection

As mentioned earlier, a hacker may try to mislead the control center to take some unnecessary action by presenting an unrealistic picture of the grid to the control center using FDIA. For example, consider the attack on a healthy system by falsification of the current signal carried out by FDIA as depicted in the Fig. 3(a). Any corrective measure based on that falsified information will critically affect the normal operation of the grid and may lead to power-cut or blackout. A natural solution to avoid the damage caused by FDIA involves securing all the sensors of the grid and that will thwart falsification of sensor information. However, the large number of sensors deployed over wide geographical span makes the task of providing security to all the individual sensors impractical because of the related financial implications [29]. A financially viable option is to secure a small set of the existing sensors. For an n bus system, assuming current and voltage monitoring at each bus, the overall sensor set τ is given as

$$\tau = [S_{I_1}, S_{V_1}, S_{I_2}, S_{V_2}, \ldots, S_{I_n}, S_{V_n}] \tag{1}$$

In the analysis of power system protection schemes, two widely used measures are referred to as *security* and *dependability*. Security refers to the ratio of the predicted no-fault cases to the actual number of no-fault cases while the dependability relates to the ratio of the detected fault cases to the actual number of faults. Now, the goal of identifying the minimal set of sensors τ_s involves finding $|\tau_s| << |\tau|$, such that the security and dependability of the overall power system protection mechanism are maintained using only the sensors from τ_s. In other words, with the information provided by τ_s, the detection of faults can be carried out. Also any sensor, which is a member of τ_s, will be protected using tamper-resistant hardware, cryptographic protocols, and encryption algorithms. Consequently, falsification of measurements transmitted from those sensors would be impossible during any FDIA.

It should be noted that the classical dimension reduction technique like PCA, which aims at reducing redundant information based on the interrelationship among different attributes is not suitable for identifying τ_s since the physical significance of individual sensor information is not maintained. The dual issues of optimally reducing the sensor information while preserving the physical significance of the data (i.e., bus voltage and current) and classification of network state (healthy/faulty) have been addressed in the present work by adopting a logical analysis of data (LAD) based classification scheme [30,31]. LAD is a data analysis technique, which uses partially defined Boolean function (pdBf) and its extensions to find patterns or rules for classification. These patterns (a.k.a. rules) can be linked to a causal-effect relationship(s) among observations and its class labels.

For the present work, observations correspond to the sensor information for a particular fault/scenario, while the class label refers to the occurrence/non-occurrence of a fault. The patterns (or rules) correlate the magnitude of current and voltage at different buses with the fault detector output i.e. 0 or 1 respectively for no fault and fault conditions. The patterns or rules generated by LAD with τ_s can be used to classify future observations, i.e., to predict the occurrence of a fault. A typical dataset comprising of different observations (power system scenarios) consists of two sets X^+ and X^- respectively comprising of sensor information during fault X^+ and no fault X^- cases.

$$X^+ = \begin{bmatrix} I_{1,1} \ V_{1,1} \ I_{2,1} \ V_{2,1} \ \ldots \ \ldots \ I_{n,1} \ V_{n,1} \\ I_{1,2} \ V_{1,2} \ I_{2,2} \ V_{2,2} \ \ldots \ \ldots \ I_{n,2} \ V_{n,2} \\ I_{1,3} \ V_{1,3} \ I_{2,3} \ V_{2,3} \ \ldots \ \ldots \ I_{n,3} \ V_{n,3} \\ . \quad . \quad . \quad . \quad . \quad . \quad . \quad . \\ . \quad . \quad . \quad . \quad . \quad . \quad . \quad . \\ . \quad . \quad . \quad . \quad . \quad . \quad . \quad . \\ I_{1,u} \ V_{1,u} \ I_{2,u} \ V_{2,u} \ \ldots \ \ldots \ I_{n,u} \ V_{n,u} \end{bmatrix} \tag{2}$$

$$\theta(X^+) = \begin{bmatrix} 1 \\ 1 \\ 1 \\ . \\ . \\ . \\ 1 \end{bmatrix} \tag{3}$$

$$X^- = \begin{bmatrix} I_{1,u+1} \ V_{1,u+1} \ I_{2,u+1} \ V_{2,u+1} \ \ldots \ \ldots \ I_{n,u+1} \ V_{n,u+1} \\ I_{1,u+2} \ V_{1,u+2} \ I_{2,u+2} \ V_{2,u+2} \ \ldots \ \ldots \ I_{n,u+2} \ V_{n,u+2} \\ I_{1,u+3} \ V_{1,u+3} \ I_{2,u+3} \ V_{2,u+3} \ \ldots \ \ldots \ I_{n,u+3} \ V_{n,u+3} \\ . \quad . \quad . \quad . \quad . \quad . \quad . \quad . \\ . \quad . \quad . \quad . \quad . \quad . \quad . \quad . \\ . \quad . \quad . \quad . \quad . \quad . \quad . \quad . \\ I_{1,m} \quad V_{1,m} \quad I_{2,m} \quad V_{2,m} \ \ldots \ \ldots \ I_{n,m} \quad V_{n,m} \end{bmatrix} \tag{4}$$

$$\theta(X^-) = \begin{bmatrix} 0 \\ 0 \\ 0 \\ . \\ . \\ . \\ 0 \end{bmatrix} \tag{5}$$

With $X^+ \cap X^- = \emptyset$.

The LAD generates positive and negative patterns corresponding to faulty and healthy scenarios from observations X^+ and X^-. The patterns are generated optimally with minimum sensor information for classifying all the cases. We refer to Subsect. 2.3 for the formal definition of a pattern.

Initially, LAD was designed to work with binary data only, in which the set of binary observations $X(= X^+ \cup X^-)$ is expressed as a pdBf ρ representing a mapping between $X \to \theta(.)\{1,0\}$. The algorithm aims at finding an approximate extension γ of ρ, such that γ can classify all the unknown observations in the sample space. In a nutshell, logical analysis of data involves the following five steps [32].

1. Binarization of Observations: For conversion of non-binary observations to binary while preserving the inherent characteristics of observations.
2. Elimination of Redundancy (or Support Sets Generation).
3. Pattern Generation.
4. Theory Formation: For Identification of a minimal set of patterns.
5. Classifier Design and Validation.

The above steps are dealt with in the subsequent sub-sections.

2.1   Binarization of Observations

For observations represented by numerical data, a threshold (a.k.a. cut-point) based method is adapted to convert the numerical data to binary. A numerical attribute β is represented in binary using two types of Boolean variables, i.e., level and interval variables. For a given cut-point c_p, we introduce a level variable $b(\beta, c_p)$ such that

$$b(\beta, c_p) = \begin{cases} 1, & \text{if } \beta \geq c_p. \\ 0, & \text{otherwise.} \end{cases} \tag{6}$$

Similarly, interval variables $b(\beta, c_p^i, c_p^j)$ are created for each pair of cut-points c_p^i and c_p^j and given by

$$b(\beta, c_p^i, c_p^j) = \begin{cases} 1, & \text{if } c_p^i \leq \beta < c_p^j. \\ 0, & \text{otherwise.} \end{cases} \tag{7}$$

The cut-point computation process is explained using an example dataset presented in the Table 1. The dataset consists of five observations with three features A, B, C. Afterward, a class label is attached to each record (Table 2). To convert the feature A to binary, a dataset is created as in the Table 3.

Further, we apply the following steps to estimate the cut-points.

1. Sort the dataset of Table 3 over A and we obtain the dataset of Table 4.
2. If two or more successive observations have identical attribute value v^i but different class labels, discard all those observations except one. Now, replace the class label of v^i by a new and unique class label. Refer to Table 5.
3. Repeat the step 2 until only unique values of the attribute are left.
4. Introduce a new cut-point $c_p^j = \frac{(A^i + A^{i+1})}{2}$, if class labels of A^i, A^{i+1} are different.

Table 1. A numerical dataset

Attributes	A	B	C
X^+: positive examples	3.5	3.8	2.8
	2.6	1.6	5.2
	1.0	2.1	3.8
X^-: negative examples	3.5	1.6	3.8
	2.3	2.1	1.0

Table 2. Dataset with class labels

A	B	C	Class Labels (Truth Values)
3.5	3.8	2.8	1
2.6	1.6	5.2	1
1.0	2.1	3.8	1
3.5	1.6	3.8	0
2.3	2.1	1.0	0

Table 3. Attribute A with class labels

A	Class Labels (Truth Values)
3.5	1
2.6	1
1.0	1
3.5	0
2.3	0

Table 4. Sorted attribute A with class labels

A	Class Labels (Truth Values)
3.5	1
3.5	0
2.6	1
2.3	0
1.0	1

Table 5. Attribute A with updated class labels

A	Class Labels
3.5	2
2.6	1
2.3	0
1.0	1

We found following cut-points using above mentioned steps.

$$c_p^1 = 3.05, \quad c_p^2 = 2.45, \quad c_p^3 = 1.65.$$

Consequently, six Boolean variables comprising of three level and three interval variables are created. After conversion of all the attributes, the binary dataset

Table 6. Binary dataset generated from the Table 2 having 15 binary variables from b_1 to b_{15}.

$A \geq 3.05$	$A \geq 2.45$	$A \geq 1.65$	$1.65 \leq A < 3.05$	$2.45 \leq A < 3.05$	$1.65 \leq A < 2.45$	$B \geq 2.95$	$B \geq 1.85$	$1.85 \leq B < 2.95$	$C \leq 4.5$	$C \leq 3.3$	$C \geq 1.9$	$1.9 \leq C < 4.5$	$1.9 \leq C < 3.3$	$3.3 \leq C < 4.5$	Class
b_1	b_2	b_3	b_4	b_5	b_6	b_7	b_8	b_9	b_{10}	b_{11}	b_{12}	b_{13}	b_{14}	b_{15}	$\mathcal{L}$
1	1	1	0	0	0	1	1	0	0	0	1	1	1	0	1
0	1	1	1	1	0	0	0	0	1	1	1	0	0	0	1
0	0	0	0	0	0	0	1	1	0	1	1	1	0	1	1
1	1	1	0	0	0	0	0	0	0	1	1	1	0	1	0
0	0	1	1	0	1	0	1	1	0	0	0	0	0	0	0

obtained is presented in the Table 6. A "categorical" attribute β can be converted into binary by associating each possible value v_i of β with a Boolean variable

$$b(\beta, v_i) = \begin{cases} 1, & \text{if } \beta = v_i. \\ 0, & \text{otherwise.} \end{cases} \tag{8}$$

2.2   Support Set Generation

Redundant attributes may be present in the binary dataset generated through binarization or any other means and removal of redundant attributes is achieved through the computation of *minimal support set*. If the projections X_M^+, X_M^- of the binary attribute set M are such that $X_M^+ \cap X_M^- = \emptyset$, then M is known as the support set of X. If removal of any constituent of M leads to $X_M^+ \cap X_M^- \neq \emptyset$, then M is known as minimal support set. For finding the minimal support set, "Mutual-Information-Greedy" algorithm from [33] has been adapted, using which the following binary features are selected.

$$M = \{b_2, b_{15}, b_8, b_1\}.$$

2.3   Modified Pattern Generation Method

In Boolean algebra, a Boolean variable or its negation is known as *literals* and conjunction of such literals is known as *term*. In LAD, if a term only covers some positive (negative) observations, then it is termed as positive (negative) *pattern*. Moreover, if a pattern is minimal, i.e., removal of any literal from the pattern leads to a pattern, which is covering both positive and negative observations, then it is called 'prime pattern'. In this paper, we have used an optimized version of the prime pattern generation technique as proposed by Boros et al. [31]. The pattern generation algorithm involves a major modification over the algorithm proposed in [31]. The modification increases the probability that the coverage of a point or observation by a single pattern only. Consequently, the 'theory formation' step used to select the most suitable pattern to cover an observation is no longer required.

After the execution of the algorithm on the projection $M = \{b_2, b_{15}, b_8, b_1\}$ of the binary dataset, following positive prime patterns are produced: (i) $\bar{b}_2 b_{15}$, (ii) $b_2 \bar{b}_{15}$. Negative prime patterns generated by following an identical procedure and the corresponding negative patterns are (i) $\bar{b}_2 \bar{b}_{15}$, (ii) $b_2 b_{15}$. It can be observed, that the binary variables appearing in the generated patterns are not dependent on the attribute B. Thus, the set of reduced attribute (or the *set of secured sensor* in the present problem) τ_s is given by $\tau_s = \{\mathbf{A}, \mathbf{C}\}$.

2.4   Design of Classifier

In this step, generated patterns are transformed into rules. Let us now consider the first positive pattern $\bar{b}_2 b_{15}$. The rule generated using the meaning of $\bar{b}_2 b_{15}$

(see Table 6) is $\neg(A \geq 2.45) \wedge (3.3 \leq C < 4.5) \implies$ 'Class label' $= 1$. One or more positive rules can be combined into 'if else-if else' structure to design a classifier (fault detector for the present problem). A simple classifier designed using the positive patterns is presented below.

Simple Classifier.

    Input: Observation consisting of attribute A, B, C.
    Output: Class label $\mathcal{L}$.
1: **if** $(\neg(A \geq 2.45) \wedge (3.3 \leq C < 4.5))$ **then**
2:    Class label $\mathcal{L} = 1$.
3: **else if** $((A \geq 2.45) \wedge \neg(3.3 \leq C < 4.5))$ **then**
4:    Class label $\mathcal{L} = 1$.
5: **else**
6:    Class label $\mathcal{L} = 0$.
7: **end if**

It can be observed from the 'Simple Classifier' also that the feature B of the original dataset is redundant and omitted by the classifier. Hence, for the present problem, the reduced sensor τ_s is given by $\tau_s = \{A, C\}$. The removal of redundant data and reduction in the number of sensors is achieved without any degradation in the classification accuracy.

3   Performance Evaluation

In this section, the efficacy of the proposed scheme in terms of optimality of τ_s, appropriateness of rules framed by LAD for fault detection and resilience against FDIA has been evaluated through comprehensive simulation studies. In this regard, the performance evaluation has been conducted on IEEE 9-bus benchmark test power system. The system has been simulated using Simulink and Simpower system toolboxes of MATLAB and executed on a 64-bit, 4 core workstation with an Intel Xeon processor and 16 GB RAM. The IEEE 9-bus system includes 9 buses, 6 lines and 3 loads as shown in Fig. 1. In the system, 54 m (3 for current and 3 for voltage measurement at each bus of the line) are deployed, which gather information at the corresponding bus.

    For generating the training dataset to derive the minimal sensor set τ_s and to frame the classification rules for fault detection based on the information from τ_s only, normal operation without attack by any adversary is considered. Normal operation incorporates scenarios associated with the healthy operation, contingencies, and faults in the power network. Observations related to healthy system state are having a class label 0. On the other hand, observations associated with a faulty system state are marked by 1 as their class label. Note that, the observations related to contingencies also have the class label as 1. The details of power system scenarios considered for training dataset preparation are presented in the Table 7.

During fault or power system contingencies (load variation and power swing) the current and voltage magnitude vary widely, and protection mechanisms present in the control center may require to take corrective measures to restore the optimal operating condition of the power system. However, if a healthy system is subjected to measures related to fault or contingencies, the consequences could be devastating. Any attacker with prior knowledge regarding the power system operation can manipulate the magnitude of voltage and current signals in order to mislead the control center to take unnecessary action whose consequence could be catastrophic.

To analyze the performance of the proposed protection scheme in terms of robustness against injection of fake data that may cause unintended operation, *test dataset* for validation has been generated by simulating several *false data injection attack* (FDIA) carried out against unsecured sensors. Two such attacks are presented in the Figs. 2(a) and 3(a). Along with FDIAs, several fault and no-fault cases have also been simulating under varying fault parameters

Table 7. Power syst. scenarios considered.

Fault parameters	Fault type	LG, 2LG, 3LG, 2L and 3L
	Fault location	1% to 100% of the line length at an interval of 5 km
	Fault inception angle	0° to 90°
	Fault resistance	0 Ω, 50 Ω and 100 Ω
Power syst. contingency	Load variation	±20%, ±40%
	Frequency variation	±2%, ±5%
	Voltage variation	±5%, ±10%
No fault		

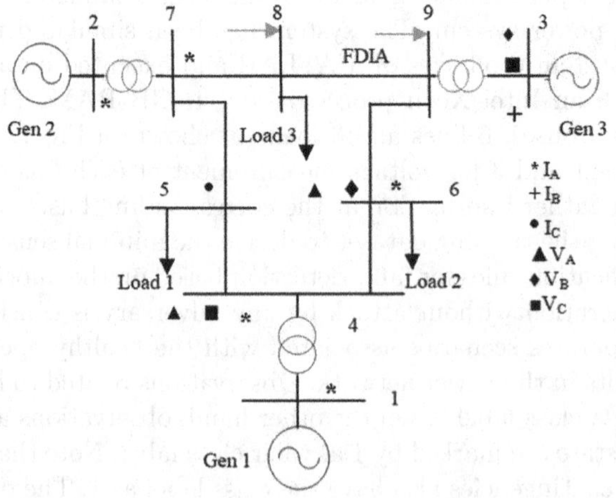

Fig. 1. IEEE 9-bus system with 13 protected sensors on different buses.

(fault location, fault inception angle and fault resistance). Some no-fault cases involving system frequency and voltage variation, switching of transmission line and sudden load encroachments have also been considered in the testing data for security analysis during healthy condition. The inclusion of FDIA, fault and no-fault cases in the test dataset allows validating the immunity of the proposed scheme against FDIA.

With the generated training dataset consisting of normal operation only, the LAD based approach discussed in Sect. 2 is employed for the design of a classifier to differentiate the healthy system state from the faulty state. The training dataset consists of 4648 observations. Among those, 4500 observations are from different faulty scenarios and rest are observations from healthy scenarios. Let us now summarize the results related to individual steps of LAD over the training set.

1. *Binarization of Observations*: In this step, 12048 binary feature variables are created from 54 current and voltage information collected from 9 different buses following the steps described in Subsect. 2.1.
2. *Support Set Generation*: Here 21 binary variables are selected from 12048 available binary variables using the method described in the Subsect. 2.2.
3. *Pattern Generation*: In this step, 28 rules are generated. The secure sensor set τ_s is also generated at this point. The details of secure sensors are available in the Fig. 1. Form the possible 54 sensors, by using only 13 secured sensors (7 currents and 6 voltages), it is possible to detect the faults. Note that, it is clear from Fig. 1 that two buses, i.e., bus-8 and bus-9 (marked by arrow) do not have any secure sensor.
4. *Classifier Design and Validation*: A classifier is built using the rules generated in the last step. The details of which are available in the Algorithm 2.

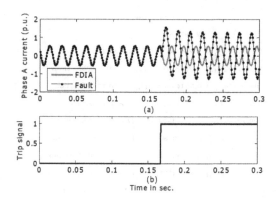

Fig. 2. (a) Suppression of current waveform of Phase "A" by FDIA during actual fault at bus-9 (b) Corresponding Trip signal by resilient protection scheme.

Let us now illustrate the results using an example. A single line to ground fault at 50 km from the bus-7 in the line between bus-7 and bus-8 has been simulated and the corresponding voltage and current waveform acquired by the unsecured sensors at bus-9 during the fault in the absence and presence of FDIA has been illustrated in Fig. 2(a). It can be observed that the current waveform in phase "A" during "AG" fault is manipulated at bus-9 by the attacker. Thus, replicating the healthy scenario post-fault, and in an attempt to mislead any fault detection process, the control center is presented with the falsified information only. The corresponding test result of the proposed FDI resilience protection scheme is shown in Fig. 2(b). It can be observed that the proposed scheme is able to detect the fault correctly even in the presence of FDIA and issued the 'Trip signal' for proper operation of the relay at the appropriate time.

Algorithm 2 Resilient Protection Scheme for IEEE 9-bus system.

1: **if** $\neg(I_{3,B} \geq 1.105100) \wedge \neg(V_{4,A} \geq 0.795880)$ **then**
2:      Fault.
3: **else if** $\neg(I_{3,B} \geq 1.105100) \wedge \neg(V_{3,C} \geq 0.722580)$ **then**
4:      Fault.
5: **else if** $(V_{4,A} \geq 0.795880) \wedge \neg(V_{3,C} \geq 0.722580)$ **then**
6:      Fault.
7: **else if** $(V_{4,A} \geq 0.795880) \wedge \neg(V_{3,B} \geq 0.744375)$ **then**
8:      Fault.
9: **else if** $(V_{4,A} \geq 0.795880) \wedge \neg(V_{4,C} \geq 0.721830)$ **then**
10:      Fault.
11: **else if** $(V_{4,A} \geq 0.795880) \wedge (I_{7,A} \geq 2.697800)$ **then**
12:      Fault.
13: **else if** $\neg(I_{5,C} \geq 0.949195) \wedge (I_{7,A} \geq 2.697800)$ **then**
14:      Fault.
15: **else if** $\neg(V_{3,C} \geq 0.722580) \wedge (V_{3,B} \geq 0.744375)$ **then**
16:      Fault.
17: **else if** $(V_{3,C} \geq 0.722580) \wedge \neg(V_{6,B} \geq 0.553810)$ **then**
18:      Fault.
19: **else if** $(V_{3,C} \geq 0.722580) \wedge (I_{7,A} \geq 2.697800)$ **then**
20:      Fault.
21: **else if** $(V_{3,C} \geq 0.722580) \wedge \neg(V_{6,A} \geq 0.567410)$ **then**
22:      Fault.
23: **else if** $(V_{3,B} \geq 0.744375) \wedge \neg(V_{4,C} \geq 0.721830)$ **then**
24:      Fault.
25: **else if** $\neg(I_{6,A} \geq 1.196150) \wedge \neg(V_{6,A} \geq 0.567410)$ **then**
26:      Fault.
27: **else if** $\neg(V_{4,C} \geq 0.721830 \wedge \neg(V_{6,A} \geq 0.567410)$ **then**
28:      Fault.
29: **else if** $\neg(I_{2,A} \geq 2.528100) \wedge \neg(V_{6,B} \geq 0.553810)$ **then**
30:      Fault.
31: **else if** $(V_{6,B} \geq 0.553810) \wedge (I_{7,A} \geq 2.697800)$ **then**
32:      Fault.
33: **else if** $(I_{3,B} \geq 1.105100) \wedge (V_{4,A} \geq 0.795880) \wedge (I_{2,A} \geq 2.528100)$ **then**
34:      Fault.
35: **else if** $(I_{3,B} \geq 1.105100) \wedge (V_{4,A} \geq 0.795880) \wedge \neg(V_{6,A} \geq 0.858385)$ **then**
36:      Fault.
37: **else if** $(I_{3,B} \geq 1.105100) \wedge (I_{6,A} \geq 0.594455) \wedge \neg(I_{6,A} \geq 0.776730)$ **then**
38:      Fault.
39: **else if** $\neg(V_{4,A} \geq 0.795880) \wedge \neg(I_{5,C} \geq 0.949195) \wedge (V_{3,C} \geq 0.722580)$ **then**
40:      Fault.
41: **else if** $\neg(V_{4,A} \geq 0.795880) \wedge \neg(I_{5,C} \geq 0.949195) \wedge \neg(I_{2,A} \geq 2.528100)$ **then**
42:      Fault.
43: **else if** $\neg(V_{4,A} \geq 0.795880) \wedge (V_{3,C} \geq 0.722580) \wedge \neg(I_{4,A} \geq 0.843595)$ **then**
44:      Fault.
45: **else if** $\neg(V_{4,A} \geq 0.795880) \wedge (V_{3,B} \geq 0.744375 \wedge \neg(I_{7,A} \geq 0.780975)$ **then**
46:      Fault.
47: **else if** $\neg(V_{4,A} \geq 0.795880) \wedge (V_{3,B} \geq 0.744375) \wedge \neg(I_{6,A} \geq 1.196150)$ **then**
48:      Fault.
49: **else if** $(I_{5,C} \geq 0.949195) \wedge \neg(I7,A \geq 0.780975) \wedge \neg(I_{4,A} \geq 0.843595)$ **then**
50:      Fault.
51: **else if** $\neg(V3,C \geq 0.722580) \wedge \neg(V4,C \geq 0.721830) \wedge (V6,B \geq 0.553810)$ **then**
52:      Fault.
53: **else if** $(V_{3,C} \geq 0.722580) \wedge (I_{7,A} \geq 0.780975) \wedge \neg(V_{4,C} \geq 0.721830)$ **then**
54:      Fault.
55: **else if** $\neg(V_{4,A} \geq 0.795880) \wedge \neg(I5,C \geq 0.949195) \wedge (V_{4,C} \geq 0.721830) \wedge \neg(I_{1,A} \geq 2.874700)$ **then**
56:      Fault.
57: **else**
58:      No Fault.
59: **end if**

Further, a healthy (no fault) case has also been analyzed in which the attacker launches an FDIA at bus-8 by replicating a single line to ground fault. The corresponding test results depicted in Fig. 3(a, b) confirm the immunity of the proposed scheme against FDIA.

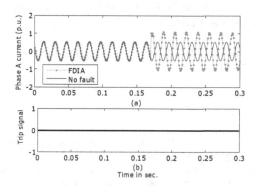

Fig. 3. (a) Suppression of current waveform of Phase "A" by FDIA during healthy operation at bus-9 (b) Corresponding Trip signal by resilient protection scheme.

Further, the performance assessment of the proposed scheme has been carried out using two statistical indices commonly used in the performance analysis of transmission line protection schemes, i.e., *dependability* and *security*. Dependability relates to the ratio of the detected fault cases to the actual number of faults while security refers to the ratio of the predicted no-fault cases to the actual number of no-fault cases. We could achieve 100% dependability and security in all the scenarios. Furthermore, the detection of fault is achieved in less than 45 μs. We have also carried out similar exercise using IEEE 39-bus bench-marked system having 234 installed voltage and current sensors, and we have achieve 100% security and dependability in this case also.

4   Conclusion

Dependence of protection algorithms on the information from the sensors spreading across wide geographical locations has increased the risk of FDIAs in power networks. In this paper, an FDIA resilient protection scheme has been proposed, in which immunity against FDIA has been achieved by securing a minimal set of sensors. The identification of sensor set contributing maximum to the system monitoring while avoiding redundancy has been carried out by employing a Boolean function based approach known as LAD. In addition of locating the strategic sensors, and thus, reducing the dimension of measured data, the LAD based approach provides a rule-based mapping between the secured sensor information, and the state (i.e., healthy or faulty) of the power system both under normal condition and FDIA. This avoids providing security to all the sensors, thereby reducing the financial cost for necessary immunity against FDIA.

The proposed computationally efficient protection scheme has been well validated for different types of faults under varying fault and power system operating parameters for IEEE three machine 9-bus system. The validation confirms the robustness of the proposed scheme against FDIA, by performing the intended relaying action. Future work in this direction is planned on extending the proposed protection scheme for fault classification, section identification, and location estimation during FDIA.

Acknowledgment. Jianying Zhou's work was supported by SUTD start-up research grant SRG-ISTD-2017-124.

References

1. Phadke, A.G., et al.: The wide world of wide-area measurement. IEEE Power Energ. Mag. **6**(5), 52–65 (2008)
2. Sridhar, S., Hahn, A., Govindarasu, M.: Cyber-physical system security for the electric power grid. Proc. IEEE **100**(1), 210–224 (2012)
3. Liang, G., Zhao, J., Luo, F., Weller, S.R., Dong, Z.Y.: A review of false data injection attacks against modern power systems. IEEE Trans. Smart Grid **8**(4), 1630–1638 (2017)
4. Liu, X., Li, Z.: False data attack models, impact analyses and defense strategies in the electricity grid. Electr. J. **30**(4), 35–42 (2017). Special Issue: Contemporary Strategies for Microgrid Operation & Control
5. Chen, K., Hu, J., He, J.: Detection and classification of transmission line faults based on unsupervised feature learning and convolutional sparse autoencoder. IEEE Trans. Smart Grid **9**(3), 1748–1758 (2018)
6. Mohanty, S.R., Pradhan, A.K., Routray, A.: A cumulative sum-based fault detector for power system relaying application. IEEE Trans. Power Deliv. **23**(1), 79–86 (2008)
7. Samantaray, S.R.: Phase-space-based fault detection in distance relaying. IEEE Trans. Power Deliv. **26**(1), 33–41 (2011)
8. Prasad, Ch.D., Nayak, P.K.: Performance assessment of swarm-assisted mean error estimation-based fault detection technique for transmission line protection. Comput. Electr. Eng. **71**, 115–128 (2018)
9. Monticelli, A.: Electric power system state estimation. Proc. IEEE **88**(2), 262–282 (2000)
10. Liu, Y., Ning, P., Reiter, M.K. False data injection attacks against state estimation in electric power grids. In: Proceedings of the 16th ACM Conference on Computer and Communications Security, CCS 2009, pp. 21–32. ACM, New York (2009)
11. Liu, Y., Ning, P., Reiter, M.K.: False data injection attacks against state estimation in electric power grids. ACM Trans. Inf. Syst. Secur. **14**(1), 13:1–13:33 (2011)
12. Deng, R., Xiao, G., Lu, R., Liang, H., Vasilakos, A.V.: False data injection on state estimation in power systems-attacks, impacts, and defense: a survey. IEEE Trans. Ind. Inf. **13**(2), 411–423 (2017)
13. Deng, R., Liang, H.: False data injection attacks with limited susceptance information and new countermeasures in smart grid. IEEE Trans. Ind. Inf. **15**(3), 1619–1628 (2019)
14. Liu, C., Wu, J., Long, C., Kundur, D.: Reactance perturbation for detecting and identifying FDI attacks in power system state estimation. IEEE J. Sel. Top. Signal Process. **12**(4), 763–776 (2018)

15. Singh, S.K., Khanna, K., Bose, R., Panigrahi, B.K., Joshi, A.: Joint-transformation-based detection of false data injection attacks in smart grid. IEEE Trans. Ind. Inf. **14**(1), 89–97 (2018)

16. Yang, L., Li, Y., Li, Z.: Improved-elm method for detecting false data attack in smart grid. Int. J. Electr. Power Energy Syst. **91**, 183–191 (2017)

17. Liu, L., Esmalifalak, M., Ding, Q., Emesih, V.A., Han, Z.: Detecting false data injection attacks on power grid by sparse optimization. IEEE Trans. Smart Grid **5**(2), 612–621 (2014)

18. Li, S., Yılmaz, Y., Wang, X.: Quickest detection of false data injection attack in wide-area smart grids. IEEE Trans. Smart Grid **6**(6), 2725–2735 (2015)

19. Yang, Q., Yang, J., Yu, W., An, D., Zhang, N., Zhao, W.: On false data-injection attacks against power system state estimation: modeling and countermeasures. IEEE Trans. Parallel Distrib. Syst. **25**(3), 717–729 (2014)

20. Bi, S., Zhang, Y.J.: Graphical methods for defense against false-data injection attacks on power system state estimation. IEEE Trans. Smart Grid **5**(3), 1216–1227 (2014)

21. Deng, R., Xiao, G., Lu, R.: Defending against false data injection attacks on power system state estimation. IEEE Trans. Ind. Inf. **13**(1), 198–207 (2017)

22. Wang, Q., Tai, W., Tang, Y., Ni, M., You, S.: A two-layer game theoretical attack-defense model for a false data injection attack against power systems. Int. J. Electr. Power Energy Syst. **104**, 169–177 (2019)

23. Yang, Q., An, D., Min, R., Yu, W., Yang, X., Zhao, W.: On optimal pmu placement-based defense against data integrity attacks in smart grid. IEEE Trans. Inf. Forensics Secur. **12**(7), 1735–1750 (2017)

24. Hao, J., Piechocki, R.J., Kaleshi, D., Chin, W.H., Fan, Z.: Sparse malicious false data injection attacks and defense mechanisms in smart grids. IEEE Trans. Ind. Inf. **11**(5), 1–12 (2015)

25. Liu, X., Li, Z., Li, Z.: Optimal protection strategy against false data injection attacks in power systems. IEEE Trans. Smart Grid **8**(4), 1802–1810 (2017)

26. Yang, Q., Jiang, L., Hao, W., Zhou, B., Yang, P., Lv, Z.: Pmu placement in electric transmission networks for reliable state estimation against false data injection attacks. IEEE Internet Things J. **4**(6), 1978–1986 (2017)

27. Ahmed, C.M., Zhou, J., Mathur, A.P.: Noise matters: using sensor and process noise fingerprint to detect stealthy cyber attacks and authenticate sensors in CPS. In: Proceedings of the 34th Annual Computer Security Applications Conference, ACSAC 2018, pp. 566–581. ACM, New York (2018)

28. Umsonst, D., Sandberg, H.: A game-theoretic approach for choosing a detector tuning under stealthy sensor data attacks. In: 2018 IEEE Conference on Decision and Control (CDC), pp. 5975–5981, December 2018

29. Kim, T.T., Poor, H.V.: Strategic protection against data injection attacks on power grids. IEEE Trans. Smart Grid **2**(2), 326–333 (2011)

30. Crama, Y., Hammer, P.L., Ibaraki, T.: Cause-effect relationships and partially defined Boolean functions. Ann. Oper. Res. **16**(1–4), 299–325 (1988)

31. Boros, E., Hammer, P.L., Ibaraki, T., Kogan, A., Mayoraz, E., Muchnik, I.: An implementation of logical analysis of data. IEEE Trans. Knowl. Data Eng. **12**(2), 292–306 (2000)

32. Alexe, G., Alexe, S., Bonates, T.O., Kogan, A.: Logical analysis of data - the vision of Peter L. Hammer. Ann. Math. Artif. Intell. **49**(1), 265–312 (2007)

33. Almuallim, H., Dietterich, T.G.: Learning Boolean concepts in the presence of many irrelevant features. Artif. Intell. **69**(1), 279–305 (1994)

Strong Leakage Resilient Encryption
by Hiding Partial Ciphertext

Jia Xu[1]([envelope]) and Jianying Zhou[2]

[1] Singtel/Trustwave, Singapore, Singapore
jiaxu2001@gmail.com
[2] Singapore University of Technology and Design, Singapore, Singapore
jianying_zhou@sutd.edu.sg

Abstract. Leakage-resilient encryption is a powerful tool to protect data confidentiality against side channel attacks. In this work, we introduce a new and strong leakage setting to counter backdoor (or Trojan horse) plus covert channel attack, by relaxing the restrictions on leakage. We allow *bounded* leakage at *anytime* and *anywhere* and over *anything*. Our leakage threshold (e.g. 10000 bits) could be much larger than typical secret key (e.g. AES key or RSA private key) size. Under such a strong leakage setting, we propose an efficient encryption scheme which is semantic secure in standard setting (i.e. without leakage) and can tolerate strong continuous leakage. We manage to construct such a secure scheme under strong leakage setting, by hiding partial (e.g. 1%) ciphertext as secure as we hide the secret key using a small amount of more secure hardware resource, so that it is almost equally difficult for any adversary to steal information regarding this well-protected partial ciphertext or the secret key. We remark that, the size of such well-protected small portion of ciphertext is chosen to be much larger than the leakage threshold. We provide concrete and practical examples of such more secure hardware resource for data communication and data storage. Furthermore, we also introduce a new notion of computational entropy, as a sort of computational version of Kolmogorov complexity. Our quantitative analysis shows that, hiding partial ciphertext is a powerful countermeasure, which enables us to achieve higher security level than existing approaches in case of backdoor plus covert channel attacks. We also show the relationship between our new notion of computational entropy and existing relevant concepts, including All-or-Nothing Transform and Exposure Resilient Function. This new computation entropy formulation may have independent interests.

Keywords: Leakage resilient encryption · Secret sharing ·
Information dispersal algorithm · Information-theoretic security ·
Side channel attack · Covert channel attack · Subliminal channel ·
Kolmogorov complexity

A full version [28] is available at https://eprint.iacr.org/2018/846.

© Springer Nature Switzerland AG 2019
J. Zhou et al. (Eds.): ACNS 2019 Workshops, LNCS 11605, pp. 172–191, 2019.
https://doi.org/10.1007/978-3-030-29729-9_10

1   Introduction

Leakage resilient cryptography has been studied for over a decade, aiming to counter side channel attacks, among other goals. Existing works on leakage resilient cryptography typically impose some restrictions on *when, where,* or *what* can be leaked. Some work assumes that there exits a leakage-free setup phase. Some works assume there exists a secure hardware device, such that any computation inside this secure device is leakage-free. If some secret key is stored in such secure device and never leaves from it, then such secret key is assumed to be leakage-free. Some works only allow leakage on secret key. Furthermore, some works consider bounded leakage with a very small upper bound—$O(\texttt{Poly}(\log \lambda))$ where λ is the security parameter.

1.1   Background in Existing Leakage Models

1.1.1   Bounded Retrieve Model

The bounded retrieve model [2,3,13,15] assumes the total amount of leaked information during the lifetime of the attacked system, is upper bounded by a constant ℓ, which could be as large as gigabytes. An existing approach [3,13] is to purposely make the shared secret key size significantly larger than the leakage upper bound—ℓ (e.g. $\geq 2\ell + \lambda$ where λ is the security parameter). In order to make the computation as fast as the case of short secret key, this approach assumes a leakage-free phase, during which, one party (say, Alice) can randomly extract a short session key from the large shared secret key using a random seed. The other party (say, Bob) of communication can re-generate the same short session key from the same shared large secret key after receiving the same random seed.

It is easy to see, under continuous bounded leakage setting, any static secret key can be leaked one bit by one bit, and pseudorandomness technique cannot be applied directly since short seed could be (partially) leaked. Furthermore, we allow $\mathcal{O}(\lambda)$ bits leakage such that leakage threshold could be larger than secret key size (e.g. the short session key in the above paragraph), thus the whole block cipher key (e.g. 128 bits AES key) could be leaked. Therefore, bounded retrieve model does not satisfy our goal.

1.1.2   A Leakage-Free Time Period During the Computation Process of Cryptography Primitive

Alwen, Dodis and Wichs [2] proposed several leakage resilient cryptography primitives with flexible (and possibly very large) key size. A key idea in their authenticated key agreement scheme, is: (1) Generate many keys in the setup; (2) and during a leakage-free time period, the sender and receiver will randomly sample a subset of keys, and use them to authenticate each other; and then establish a short shared session key. As long as a constant fraction of all keys are unknown to the adversary after bounded leakage, a random subset of keys contains at least one unknown key with very high probability. After that, standard cryptography primitives are applied with the short secure session key (e.g. AES).

In our leakage setting, there will be *no* leakage-free time period and any *short* value (e.g. AES key) could be leaked. So we have to seek new approaches.

1.1.3   Secret Key Never Leaves from Secure Hardware Device

The computation power of secure hardware devices (e.g. Trusted Platform Module) may not be able to match the power of desktop Intel/AMD CPU. Furthermore, there seems no evidence to show that the vendors of secure hardware device are more trusted than vendors of other component (e.g. CPU, GPU, RAM, hard disk, OS, web browser, virtual machine software, etc.) in a computer system.

1.1.4   Randomness Extractor

One may consider to extract a short block cipher (e.g. AES) key from a long secret key and then encrypt the message using the short block cipher directly. Assuming leakage only occurred before the randomness extractor was applied, (e.g. as the setting of [3, 13]), this method will work. But in our setting, we do not make such assumption, and instead we allow bounded leakage at any time.

1.1.5   Memoryless Leakage Oracle

An essential difference between leakage oracle in side channel attack in related works and leakage oracle in Trojan horse malware plus covert channel attack in this paper, is that, whether the leakage oracle has cache memory and is allowed to access history data. Some recent works in leakage resilient cryptography [6, 7, 23] assumes that: (1) for each invocation of cryptography primitive, the leakage threshold is smaller than secret key size; and (2) leakage oracle only takes input from current status of the cryptography computation, and is not allowed to access historical status. They can achieve security by refreshing the secret key frequently (together with other techniques). Imagine a simplified example [23]: To encrypt the i-th message, one may adopt a fresh 256-bit encryption key $k_i := \mathrm{SHA256}(k_{i-1})$, and the adversary is allowed to learn only a single bit $\mathcal{L}(k_i) \in \{0,1\}$ over the key k_i. With all leaked information $\{\mathcal{L}(k_j) : j \in [0,i]\}$, a polynomial-time adversary seems not be able to learn some useful knowledge about any secret key. However, in case of Trojan horse plus covert channel attack in this paper, the Trojan horse malware may keep an old key k_0 in a local cache memory, and send out one bit per every invocation of encryption scheme via covert channel. So after encrypting $|k| = 256$ messages, all of 256 bits of k_0 could be sent out to a remote adversary, who can compute every k_i from k_0. With all ciphertexts (which can be obtained via eavesdropping, without resorting to leakage oracle), the adversary can decrypt and recover all plaintexts. Thus 256 bits leakage leads to exposure of everything—all plaintexts and (future) secret keys. Our new security formulation in this work is aiming to prevent such kind of *leakage amplification*.

It will be interesting to study the leakage resilient cryptography with adversary who has limited leakage bandwidth (say ℓ bits per invocation of crypto primitive) and limited cache memory (say w bits memory). In this work, we actually do not assume any upper bound in the size of cache memory. Since covert channel with large bandwidth and/or Trojan horse with large cache memory, may be more easily captured or prevented by existing solution (e.g. anti-virus software and intrusion detection system, Trojan-Resilient hardware [9,16]), it is reasonable to put some small upper bound in values of ℓ and w. We leave this as an open problem.

1.2 Our Contributions

The main contributions of this work can be summarized as below.

1.2.1 New Leakage Setting

Since existing leakage settings does not fit for our goal, we present a new strong leakage model, to capture the threat of backdoor or Trojan horse and covert channels in computer hardware/software systems. We allow *bounded* (e.g. 10000 bits) leakage at *anytime* and *anywhere* and over *anything*, with only two restrictions on the adversary: (1) the adversary algorithms are efficient (probabilistic polynomial time); (2) the bandwidth of the covert channel is bounded from the above. By our knowledge, all existing works designed for leakage settings in Sect. 1.1 are trivially broken under our leakage setting, since the Trojan horse could observe every step of computation of the victim program (e.g. an encryption program) and then steal the entire short private key. We emphasize that, the white box cryptography [5,18] using program obfuscation, which claims to protect secret key from attackers with direct control of the encryption device, is prohibitively impractical, even for a simple function [12].

1.2.2 Notion of Steal-Entropy

We propose a new notion called "steal-entropy", as a sort of computational version of Kolmogorov complexity. With this "steal-entropy", we quantitatively analyse the advantage of our approach over existing works. Our formulation is non-trivial and has to resolve several important issues: (1) Unlike Shannon-Entropy, Yao-Entropy and Hill-Entropy are defined over distribution of random variable, and Kolmogorov complexity is defined over string, our steal-entropy will be defined over an algorithm which converts the distribution of input random variable to the distribution of output random variable. (2) Statistical or computational indistinguishability notion (e.g. semantic security under CPA/CCA/CCA2 attack mode) is inappropriate in our leakage setting, since a single bit of arbitrary leakage will help an adversary to win the guess-game trivially. (3) Kolmogorov complexity is uncomputable in general, but in our formulation, we should avoid to define any uncomputable function. As a result, unlike existing variant formulations of entropy, it is hard to define our steal-entropy as a single scalar value (More discussion is available in our full version).

Instead, we will give an upper bound and a lower bound for the steal-entropy of a given algorithm. To show a program has poor steal-entropy, we need provide a small upper bound on the steal-entropy of this program; to show a program has high steal-entropy, we need provide a large lower bound on the steal-entropy of this program.

1.2.3 Construction

We propose an efficient encryption scheme and demonstrate that hiding partial ciphertext could be a powerful tool to defeat strong leakage attack. We construct our encryption scheme using Vandermonde matrix and evaluate the steal-entropy of the proposed scheme without relying on any hard problem assumption. Informally speaking, our encryption scheme will ensure that, without complete ciphertext, the attacker obtains very limited information about the plaintext, even if the attacker has stolen a bounded amount of message (e.g. the entire short private key) of his/her choice. We will compare our solution with some related approaches, including All-or-Nothing Transform and White-Box Cryptography, both of which could not satisfy our goal.

The proposed solution will be used to construct a "virtually isolated network" [29]. We discuss details later in Sect. 2.

1.3 Organizations

The rest of this paper is organized in this way: Sect. 2 gives an overview of our work, including our leakage setting, formulation of steal-entropy, and our proposed construction of leakage/steal-resilient encryption scheme. In addition to the related works already discussed in Sects. 1 and 2, Sect. 3 discusses more related works. We present our formal formulation of steal-entropy in Sect. 4, propose and analyse our encryption scheme in Sect. 5. We conclude this paper in Sect. 6. A full version with more details is available online [28].

2 Overview of Our Work

2.1 Our Leakage Setting

2.1.1 Motivation of New Leakage Setting

In this paper, we aim to counter not only side channel attack but also covert channel attack. Nowadays, computer systems become so complex and consist of a lot of software/hardware components which are designed, manufactured and sold by various companies from various countries. It is definitely not a trivial task for PC users to check whether some backdoor program or malware (e.g. Trojan horse) has been planted inside his/her PC hardware/software system. The well-known "Dual Elliptic Curve Deterministic Random Bit Generator" (Dual_EC_DRBG) backdoor[1] demonstrates that the potential threat from backdoor is not that far away from every computer user. Another serious threat is

---

[1] https://en.wikipedia.org/wiki/Kleptography and https://en.wikipedia.org/wiki/Dual_EC_DRBG.

software Trojans horse or even hardware Trojan horse[2]. The backdoor or Trojan horse malware may observe the victim's computer system to gather information and send collected (possibly compressed) information out via a covert channel or subliminal channel.

Facing such threats from backdoor and Trojan horse, in this work, we have to revise the existing leakage setting: (1) Theoretically, backdoor or Trojan horse programs could be planted by some software/hardware vendor and they exist in victim's computer from the very beginning. So it might not be appropriate to assume a leakage-free time period. (2) Possibly, the backdoor program might be planted by vendors of the secure hardware device and the assumption of leakage-free secure hardware device is hard to validate. (3) The backdoor or Trojan horse malware may have their own storage buffers, so history data can be buffered and then leaked 1 bit by 1 bit via the covert channel (thus Pereira, Standaert and Vivek [23] would be broken trivially as discussed in Sect. 1.1.5).

2.1.2   New Leakage Setting

In general, we allow *efficient* leakage with *bounded bandwidth* at *anytime* and *anywhere* and over *anything*. The only two restrictions on leakage are: (1) The leakage amount of each encryption (i.e. the bandwidth of covert channel) is bounded (e.g. $\mathcal{O}(\lambda)$). In this paper, we are interested in medium value of leakage threshold, e.g. tens of thousands bits, which is much larger than typical private key size (e.g. AES key and RSA private key). (2) The backdoor or Trojan horse program (i.e. the leakage function) is computationally bounded (e.g. polynomial time algorithm). Our setting is closer to study of memory leakage resilient cryptography, and does not follow the assumption that *only computation leaks information* [22].

Recall that, in most, if not all, leakage-resilient cryptography research works, an adversary has two different methods to obtain desired information:

- A *cheap* method to obtain a large amount of weakly protected information, for example, eavesdropping ciphertext on communication link.
- An *expensive* method to obtain a small amount of strongly protected information, for example, using side channel attack or Trojan horse malware plus covert channel attack to obtain partial or full information of the short secret key.

Typically in existing works, an adversary is assumed to obtain full information of ciphertext using the cheap method (e.g. eavesdropping), meanwhile subject to several restrictions on obtaining information of short secret key (e.g. assumed leakage-free time period or hardware device). Unlike existing works, in this paper, we impose minimum restrictions on information leakage, and assume that a small part (e.g. 1% or 0.1%) of ciphertext[3] is as strongly protected as the short secret

---

[2] http://spectrum.ieee.org/semiconductors/design/stopping-hardware-Trojans-in-their-tracks.

[3] The encryption scheme is length-preserving, and the size of ciphertext is equal to the size of plaintext.

key, so that the adversary has to resort to the expensive method (e.g. Trojan horse and covert channel) to obtain this part of ciphertext. Next, we will support this assumption with real world examples.

Secure Storage Device. For data storage, we assume there are two categories of storage: one with small capacity is relatively more expensive, in term of unit price, but much more secure; the other with large capacity is cheaper but insecure. In case that a user wish to backup large size sensitive historical data in cloud storage server, but did not trust the cloud in data confidentiality. Then this user's local *offline* storage device, which is physically disconnected from any computers and Internet, could be an example of the former, and the cloud storage[4] could be an example of the latter.

Secure Communication Link. For data transmission, we assume there exist two categories of communication channels, one with small bandwidth is very expensive but much more secure, such that an adversary cannot obtain the transmitted data with low cost (e.g. eavesdropping); the other with large bandwidth is cheap but insecure, such that an adversary can obtain all transmitted data with low cost. The example of former could be satellite link (or even neutrinos communication in the future), which is relatively more difficult to eavesdrop, and the example of latter could be Internet. Another example is "virtually isolated network"[5], recently proposed by Xu and Zhou [29], which is a hybrid network with two communication channels: one is a physically isolated network with small bandwidth, and the other is Internet with large bandwidth. Their work [29] combines these two channels with unidirectional network links (a.k.a data diode or air gap), so that the isolated network will be still always physically isolated from Internet.

Our strategy is to enhance security level of the large amount of cheap but insecure hardware resource by leveraging on small amount of expensive but more secure hardware resource, essentially creating a hybrid effects in security. We aim to prevent the adversary from eavesdropping full information of our ciphertext.

2.2   Notion of Steal-Entropy

Unlike previous leakage formulation, we attempt to formalize security in leakage setting from a different angle. We try to answer a very important question:

"At least how many bits should the adversary steal in order to obtain the desired secret information?"

---

[4] Note: (1) Many cloud storage servers provide a certain amount (e.g. 15GB) of free cloud storage for individual users; (2) the cost of offline local storage should include not only hardware purchase cost but also hardware maintenance and storage cost (i.e. keep the harddisk drive in a proper physical environment for a long time).

[5] Actually, the motivation of this work is to provide an extremely secure (*informally, close to physically isolated network*) communication method in this "virtually isolated network" [29]. Here we choose strong leakage resilience against potential backdoor as our formal definition of "extremely secure".

In this work, we are concerning how many bits the adversary has to obtain *using the expensive method*, in order to obtain full or partial information of the plaintext. Informally, we may call this "minimum but sufficient number of leaked/stolen bits" which will lead to compromise of secret plaintext, as the *steal-entropy* of the encryption algorithm.

Let P (e.g encryption algorithm/program) denote the victim algorithm or program. In our formulation, an adversary chooses two algorithms, denoted with steal algorithm S and recovery algorithm R. The steal algorithm S is given oracle access to the whole computation process of P, including any internal states (e.g. secret keys, random seeds, input and any computation steps). Then the steal algorithm S is allowed to pass a short message, which is at most ℓ bits, to the recovery algorithm R, which attempts to output desired secret information. If the recovery algorithm R is able to output the desired secret information with probability close to 1, with value of ℓ much smaller than the size of desired secret information, then we say the victim algorithm P has very low steal-entropy rate. In this work, we are interested in medium value of leakage threshold ℓ (e.g. tens of thousands), which is larger than typical secret key length, but could be much smaller than typical ciphertext length. Our notion of "steal-entropy" could be treated as a computation version of Kolmogorov complexity.

2.2.1   Steal-Entropy in Input or Output

Pseudorandom number generators, pseudorandom function and encryption are important cryptography primitives applied to protect data confidentiality. For an algorithm P similar to pseudorandom number generator and pseudorandom function, we are interested to ask a question: Assuming a Trojan horse malware is observing the computation process of algorithm P upon a randomly chosen input x, at least how many bits should the Trojan horse malware steal and send out, in order to allow a remote attacker to recover the output $P(x)$ of the algorithm P? To address this question, we define a notion called "Steal-Entropy of an algorithm in Output". Due to space constrain, we will leave the formal definition of this notion in the full version of this paper.

For algorithm P similar to encryption scheme, we are interested to ask another question: Assuming a Trojan horse malware is observing the computation process of algorithm P upon a randomly chosen input x, at least how many bits should this Trojan horse malware steal and send out, in order to allow a remote attacker to recover the input x, where this remote attacker has access to the output[6] $P(x)$? To address this question, we define a notion called "Steal-Entropy of an algorithm in Input". In addition, to deal with partial information protection, we define a notion called "Strong Steal-Entropy of an algorithm".

2.2.2   Relation with Existing Similar Notions

We also formally analyze the differences between our notion of steal-entropy with existing similar notions, including Yao-Entropy [30], Hill-Entropy [19],

---

[6] Usually, it is assumed that the adversary has access to the ciphertext.

Information Dispersal Algorithm [24], All-or-Nothing Transform [25], and Exposure Resilient Function [10]. We manage to separate our proposed steal-entropy from all of these existing formulations. More details are in our full version [28].

2.3   Our Approach

When the leakage threshold ℓ is larger than typical secret key size, most existing encryption schemes and leakage resilient encryption schemes (which only tolerates leakage upto $O(poly \log \lambda) < \lambda$ bits, where λ is the security parameter) would fail to protect data confidentiality, since in typical setting, an adversary could obtain all ciphertext with low cost (e.g. eavesdropping), and the secret decryption key could be stolen by Trojan horse malware and delivered to the remote adversary via covert channel.

Facing such stringent threat of medium size of arbitrary information leakage, two possible directions are: (1) Construct novel encryption scheme with larger flexible key size, say the encryption/decryption key size could be a user-tunable parameter, and range from hundreds bits to hundreds of thousands bits or even more. We will report our work in this direction in a separate paper. We remark that Alwen, Dodis and Wichs [2] does not satisfy our purpose, since this work [2] eventually extracted a short session key from arbitrary large size long term secret key, where this extracted short session key could be stolen under our leakage setting. (2) Break the assumption that the adversary could *easily* obtain all ciphertext. Indeed, this work will attempt to hide a small portion of ciphertext using more secure hardware resource, so that the adversary has to resort to the expensive method to steal information about this small portion of ciphertext, in a similar way that he/she steals the secret key.

2.3.1   Randomness Source

Any static secret information might be stolen one bit by one bit, if backdoor or Trojan horse exists. To defeat continuous leakage/steal with buffer storage, we have to keep investing more and more randomness. However, it is expensive to generate cryptographically secure randomness. In our solution, we will exploit the fact that *plaintext itself is naturally a sort of random source to the view of adversary*, saving the cost to generate true randomness. We protect a small portion of the ciphertext using more secure hardware resource, so that this portion of ciphertext actually acts as another "secret key", which is derived from the plaintext and will change naturally with plaintext, to the view of adversary.

2.4   Our Construction

Our leakage setting provides much more freedom and power to adversary, compared to existing works on leakage-resilient cryptography. Consequently, the two very important classical tools, namely *computational indistinguishability* and (statistical or computational) *randomness extractor*, are hardly to be applied under our formulation. In this work, we have to resort to information theory techniques.

Definition 1 (Blockwise Uniform Distribution). *Let* $\mathbf{y} = (\mathbf{y}_1, \mathbf{y}_2, \cdots, \mathbf{y}_n)$, *where* $\mathbf{y}_i \in \{0,1\}^\rho$ *for each* $i \in [1, n]$. *We say* $\mathbf{y}$ *follows* (ζ, ρ)-*Blockwise-Uniform Distribution, if for any subset* $S = \{i_1, i_2, \cdots, i_\zeta\} \subset [1, n]$ *with* $|S| = \zeta$ *and* $i_1 < i_2 < i_3 < \cdots < i_\zeta$, *we have the joint Shannon-entropy*

$$\mathbb{H}^{\text{Shannon}}(\mathbf{y}_{i_1}, \mathbf{y}_{i_2}, \cdots, \mathbf{y}_{i_\zeta}) = \rho\zeta. \tag{1}$$

That is, any subset of ζ *distinct blocks* $\mathbf{y}_i$ *will have joint Shannon entropy equal to their total bit-length (i.e. entropy rate equal to 1).*

Remark 1. When $\rho = 1$ and $\zeta = n$, then (ζ, ρ)-Blockwise-Uniform Distribution is identical with uniform distribution.

In this work, we will construct an invertible algorithm P using Vandermonde matrix, such that its inverse algorithm P^{-1}, satisfies this property:

Property 1. *Let* Ctx_0 *and* Ctx_1 *be the small share and large share of cipher-text, and assume the bit-length* $|\mathsf{Ctx}_1| = \tau \cdot |\mathsf{Ctx}0| = \tau\rho\zeta$. *If* Ctx_0 *is independently and uniformly randomly distributed over* $\{0,1\}^{\rho\zeta}$, *then the output* $x = \mathsf{P}^{-1}(\mathsf{Ctx}_0, \mathsf{Ctx}_1)$ *follows* (ζ, ρ)-*Blockwise-Uniform Distribution, regardless of value of* Ctx_1 *(e.g. this value could be fixed to any given bit-string from its domain).*

Suppose somehow an attacker is able to output ζ bits among x_i's, say x_{i_j}, $j \in [1, \zeta]$. Then these ζ bits x_{i_j}'s will reside in at most ζ distinct ρ-bit blocks in bit-string x. Since any subset of ζ blocks of x will have Shannon entropy rate equal to 1 (i.e. entropy equal to the bit-length), the collection of these ζ bits x_{i_j}'s will have exactly ζ bits Shannon entropy. Therefore, the adversary has to steal at least ζ bits message via the covert channel, as desired. Apparent, the above proof is not tight with a multiplicative loss of factor ρ. We leaf the tight proof with better security parameters in future work.

3 Related Works

The related works in leakage resilient cryptography have been discussed in Sect. 1.1. Here we discuss other related works.

Symmetric encryption scheme (e.g. AES, Blowfish[7], and Triple DES[8].) could be the most widely adopted cryptographically secure primitive to protect data confidentiality, especially for large volume of data. AES [11] is a typical example of symmetric encryption scheme, and has been actively adopted in industry and research area due to its security and efficiency for more than one decade.

In additional to encryption techniques, another well-known cryptographic primitive that can be used to protect data confidentiality is "secret-sharing" scheme invented by Shamir [26]. Compared to encryption scheme (e.g. AES [11])

---

[7] https://www.schneier.com/academic/blowfish/.
[8] http://csrc.nist.gov/publications/nistpubs/800-67-Rev1/SP-800-67-Rev1.pdf.

which can only achieve conditional security, secret-sharing scheme may achieve unconditional security (also known as information-theoretic security), assuming the adversary cannot collect sufficient number of shares.

Despite its strong security, Shamir's secret sharing scheme has significant drawbacks when protecting data confidentiality: (1) for (t, n)-secret sharing scheme, the storage overhead is as large as $(n-1)$ times of size of the secret (i.e. the plaintext to be protected); (2) the reconstruction [21] (or decoding) process is not as efficient as DES or AES.

Rabin [24] proposed "information dispersal algorithm" with zero storage overhead, such that the sum of sizes of all shares is equal to the size of secret message size. His solution is conceptually simple: Let row vector $m = (m_0, m_1, \ldots, m_n)$ be the secret message. Choose an invertible n by n matrix $\mathbf{T}$ with inverse matrix $\mathbf{T}^{-1}$. By multiplying row vector m with matrix $\mathbf{T}$, we obtain the n shares $c = (c_0, c_1, \ldots, c_{n-1}) = m \times \mathbf{T}$. Accordingly, the original secret message m can be recovered by matrix multiplication $m = c \times \mathbf{T}^{-1}$. Othman and Mokdad [8] proposed to protect communication confidentiality by sending each share of message in distinct network path from the same sender to the same receiver.

Alternatively, Krawczyk [20] attempted to make each share shortened, by dividing ciphertext of the long secret message into n pieces, and then apply Shamir's secret sharing scheme over the encryption key. Thus, the storage overhead is linear in short encryption key size and is a fraction of secret message size.

4   Steal-Entropy: How Many Bits Should Be Stolen to Recover the Secrete Information?

In this section, we propose the notion of "Steal-Entropy". Unlike traditional entropy concepts (e.g. Shannon-Entropy, Yao-Entropy[9], Hill-Entropy, etc) which are defined over random variable with a certain distributions, "steal-entropy" will be defined over algorithms which convert input distribution to output distribution. Our notion of "steal-entropy" could be considered as a computational version of Kolmogorov Complexity [4], which is quoted in full version.

4.1   Steal-Entropy of an Algorithm in Input

Definition 2 (Steal-Entropy of an Algorithm in Input). *Let* $\mathsf{P} : \{0,1\}^n \to \{0,1\}^m$ *be a deterministic[10] single-input algorithm. Let* $\epsilon \in [0, \frac{1}{4})$. *Let* $\mathcal{A}$ *be a* t-*adversary associated with a pair of algorithms* (S, R), *such that*

- *both the steal (or stealage) algorithm* S *and the recovery algorithm* R *are probabilistic algorithms within time* t, *and*

---

[9] Shannon-Entropy is information-theoretical. Both Yao-Entropy and Hill-Entropy are computational variants.

[10] When all random coins are treated as a part of input, any probabilistic algorithm will become deterministic.

- *for any non-negative integer ℓ, the steal algorithm*

$$\mathsf{S}^{\mathcal{O}(\mathsf{P}(x))}(\ell) \in \{0,1\}^{\leq \ell} \setminus \{EmptyString\}$$

with oracle access to P, is allowed to observe all internal states during computation process of algorithm P upon an input x, and outputs at most ℓ bits non-empty steal-message, and
- *the recovery algorithm R takes as input the value $\mathsf{P}(x)$ and the steal-message generated by $\mathsf{S}(\ell)$, and attempts to guess the value x.*

We make the following definitions.

- *We define the advantage of $\mathcal{A}$ against P w.r.t. input $x \in \{0,1\}^n$ as below*

$$\mathsf{Adv}_{\mathcal{A}(\ell),\mathsf{P}}^{\mathrm{in}}(x) = \Pr\left[\mathsf{R}\left(\mathsf{S}^{\mathcal{O}(y \leftarrow \mathsf{P}(x))}(\ell), y\right) = x\right] \tag{2}$$

where the probability is taken over all random coins of algorithms S and R.
- *We say the **infimum of Steal-Entropy in Input of algorithm** P is at least ξ, denoted as $\inf \mathbb{S}_{\epsilon,t}^{\mathrm{in}}(\mathsf{P}) \geq \xi$, if for any t-adversary $\mathcal{A}$, for any non-negative integer $\ell \leq \xi$,*

$$\Pr_{x \xleftarrow{R} \{0,1\}^n}\left[\mathsf{Adv}_{\mathcal{A}(\ell),\mathsf{P}}^{\mathrm{in}}(x) \leq \frac{1}{2^{\xi-\ell}} + \epsilon\right] \geq 1 - \epsilon. \tag{3}$$

- *We say the **supremum of Steal-Entropy in Input of algorithm** P is at most ξ, denoted as $\sup \mathbb{S}_{\epsilon,t}^{\mathrm{in}}(\mathsf{P}) \leq \xi$, if for some t-adversary $\mathcal{A}$,*

$$\Pr_{x \xleftarrow{R} \{0,1\}^n}\left[\mathsf{Adv}_{\mathcal{A}(\xi),\mathsf{P}}^{\mathrm{in}}(x) \geq 1 - \epsilon\right] \geq 1 - \epsilon. \tag{4}$$

- *We say $\mathbb{S}_{\epsilon,t}^{\mathrm{in}}(\mathsf{P}_0) \geq \mathbb{S}_{\epsilon,t}^{\mathrm{in}}(\mathsf{P}_1)$ (or equivalently $\mathbb{S}_{\epsilon,t}^{\mathrm{in}}(\mathsf{P}_1) \leq \mathbb{S}_{\epsilon,t}^{\mathrm{in}}(\mathsf{P}_0)$), if the following two equations hold*

$$\inf \mathbb{S}_{\epsilon,t}^{\mathrm{in}}(\mathsf{P}_0) \geq \inf \mathbb{S}_{\epsilon,t}^{\mathrm{in}}(\mathsf{P}_1); \qquad \sup \mathbb{S}_{\epsilon,t}^{\mathrm{in}}(\mathsf{P}_0) \geq \sup \mathbb{S}_{\epsilon,t}^{\mathrm{in}}(\mathsf{P}_1). \tag{5}$$

- *We say $\mathbb{S}_{\epsilon,t}^{\mathrm{in}}(\mathsf{P}_0) \gg \mathbb{S}_{\epsilon,t}^{\mathrm{in}}(\mathsf{P}_1)$ (or equivalently, $\mathbb{S}_{\epsilon,t}^{\mathrm{in}}(\mathsf{P}_1) \ll \mathbb{S}_{\epsilon,t}^{\mathrm{in}}(\mathsf{P}_0)$), if the following equation holds*

$$\inf \mathbb{S}_{\epsilon,t}^{\mathrm{in}}(\mathsf{P}_0) \geq \sup \mathbb{S}_{\epsilon,t}^{\mathrm{in}}(\mathsf{P}_1). \tag{6}$$

Proposition 1. *If P is an invertible algorithm, and the inverse algorithm P^{-1} has running time $\leq t$, then $\inf \mathbb{S}_{\epsilon,t}^{\mathrm{in}}(\mathsf{P}) = \sup \mathbb{S}_{\epsilon,t}^{\mathrm{in}}(\mathsf{P}) = 0$.*

When the encryption/decryption key is fixed, an encryption algorithm Enc is an invertible algorithm from plaintext to ciphertext. Before any information leakage, an adversary may have knowledge of the whole family $\{\mathsf{Enc}_k\}_{k \leftarrow \mathsf{KGen}(1^\lambda)}$ and do not know which one is picked from this family of permutation algorithms. By stealing the key k, an adversary is able to recover plaintext from ciphertext. This simple fact is summarized as below.

Proposition 2. *For any PPT encryption scheme $(\mathsf{KGen}, \mathsf{Enc}, \mathsf{Dec})$ and for any key k generated by KGen, we have $\sup \mathbb{S}_{\epsilon,t}^{\mathrm{in}}(\mathsf{Enc}_k) \leq |k|$, where $\epsilon = 0$, and $t = \mathrm{poly}(\cdot)$.*

4.2 Discussion

An interesting question is to evaluate the steal-entropy for classical hard problems: factorization problem and discrete log problem, where thousands (say 2048) bits long key provides roughly 80 bits security level. $\mathsf{P_{Fact}}(p, q) = p \times q$ where both p and q are primes with equal bit-length. $\mathsf{P_{Log}}(x) = g^x \mod p$ where both g and p are public constants, p is a prime and g is a generator modulo p. Will the steal-entropy of these algorithm be closer to their key size (i.e. thousands) or security level (i.e. 80)? We leave it as an open problem.

4.3 Strong Steal-Entropy in Input

Informally, after stealing ℓ bits arbitrary message, the adversary should be unable to output $\ell + \Delta$ bits information about the secret value, and there will be no leakage amplification.

Definition 3 (Strong Steal-Entropy of an Algorithm in Input). *Let* P : $\{0,1\}^n \rightarrow \{0,1\}^m$ *be a deterministic[11] single-input algorithm. Let* $\epsilon \in [0, \frac{1}{4})$. *Let* $\mathcal{A}$ *be a t-adversary associated with a pair of algorithms* (S, R), *such that*

- *both the steal (or stealage) algorithm* S *and the recovery algorithm* R *are probabilistic algorithms within time* t, *and*
- *for any non-negative integer* ℓ, *the steal algorithm*

$$\mathsf{S}^{\mathcal{O}(\mathsf{P}(x))}(\ell) \in \{0,1\}^{\leq \ell} \setminus \{\textit{EmptyString}\}$$

  with oracle access to P, *is allowed to observe all internal states during computation process of algorithm* P *upon an input* x, *and outputs at most* ℓ *bits non-empty steal-message, and*
- *the recovery algorithm* R *takes 2 inputs: (1) the steal-message generated by* $\mathsf{S}(\ell)$, *and (2) the value* $\mathsf{P}(x)$, *and outputs two values: (1)* $\bar{x} \in \{0,1\}^n$, *which is a guess of* x, *and (2) a subset of indices* $\mathbf{I}_x \subset [1, n]$.

We introduce the following definitions.

- *For any adversary* $\mathcal{A}$ *with steal algorithm* S *and recovery algorithm* R, *let us define the set* $\mathbf{G}_{\mathrm{msg}}$ *of good steal-message as below*

$$\mathbf{G}_{\mathrm{msg}}^{\mathsf{R}}(\ell, \Delta, x, \beta) \overset{\text{def}}{=} \left\{ \mathsf{Msg} \in \{0,1\}^{\leq \ell} : \begin{array}{l} (\bar{x}, \mathbf{I}) \leftarrow \mathsf{R}(\mathsf{Msg}, \mathsf{P}(x)); \\ |\mathbf{I}| \geq \ell + \Delta; \\ \forall i \in \mathbf{I}, \Pr[\bar{x}[i] = x[i]] \geq \beta \end{array} \right\} \tag{7}$$

  where the probability is taken over the random coins of R.

---

[11] When all random coins are treated as a part of input, any probabilistic algorithm will become deterministic.

- *Similarly, let us define the set $\mathbf{G}_x$ of good input x as below*

$$\mathbf{G}_x^{\mathsf{S},\mathsf{R}}(\ell,\Delta,\alpha,\beta) \stackrel{\text{def}}{=} \left\{ x \in \{0,1\}^n : \Pr[\mathsf{S}^{\mathcal{O}(\mathsf{P}(x))}(\cdot) \in \mathbf{G}_{\mathrm{msg}}^{\mathsf{R}}(\ell,\Delta,x,\beta)] \ge \alpha \right\} \tag{8}$$

where the probability is taken over the random coins of S.
- *We say the **supremum of Strong Steal-Entropy in Input of algorithm** P is at most ξ, denoted as $\sup \mathbb{S}_{\epsilon,t}^{\mathrm{sin}}(\mathsf{P}) \le \xi$, if for some t-adversary $\mathcal{A} = (\mathsf{S},\mathsf{R})$,*

$$\Pr_{x \in_R \{0,1\}^n} [x \in \mathbf{G}_x^{\mathsf{S},\mathsf{R}}(\xi, \varsigma(\xi,\epsilon)+1-\ell, 1-\epsilon, 1-\epsilon)] \ge 1-\epsilon \tag{9}$$

where function $\varsigma(\cdot,\cdot)$ is defined as below[12]

$$\varsigma(\ell,\epsilon) \stackrel{\text{def}}{=} \begin{cases} \ell, & \text{if } 0 \le \epsilon < 2^{-(\ell-1)} \\ \ell+1, & \text{if } 2^{-(\ell-1)} \le \epsilon < \frac{1}{4}. \end{cases} \tag{10}$$

- *Let $\epsilon \ge \lambda^{-c}$ where c could be any positive integer. We say the **infimum of Strong Steal-Entropy in Input of algorithm** P is at least ξ, denoted as $\inf \mathbb{S}_{\epsilon,t}^{\mathrm{sin}}(\mathsf{P}) \ge \xi$, if for any t-adversary $\mathcal{A} = (\mathsf{S},\mathsf{R})$, for any ℓ with $\varsigma(\ell,\epsilon) = \ell+1 < \xi$,*

$$\Pr_{x \in_R \{0,1\}^n} [x \in \mathbf{G}_x^{\mathsf{S},\mathsf{R}}(\ell, \varsigma(\ell,\epsilon)+1-\ell, 0.5+\epsilon, 0.5+\epsilon)] \le 0.5 + negl(\lambda), \tag{11}$$

where λ is the security parameter, and $negl(\cdot)$ denotes some negligible function.
- *We say $\mathbb{S}_{\epsilon,t}^{\mathrm{sin}}(\mathsf{P}_0) \ge \mathbb{S}_{\epsilon,t}^{\mathrm{sin}}(\mathsf{P}_1)$ (or equivalently $\mathbb{S}_{\epsilon,t}^{\mathrm{sin}}(\mathsf{P}_1) \le \mathbb{S}_{\epsilon,t}^{\mathrm{sin}}(\mathsf{P}_0)$), if the following two equations hold*

$$\inf \mathbb{S}_{\epsilon,t}^{\mathrm{sin}}(\mathsf{P}_0) \ge \inf \mathbb{S}_{\epsilon,t}^{\mathrm{sin}}(\mathsf{P}_1); \qquad \sup \mathbb{S}_{\epsilon,t}^{\mathrm{sin}}(\mathsf{P}_0) \ge \sup \mathbb{S}_{\epsilon,t}^{\mathrm{sin}}(\mathsf{P}_1). \tag{12}$$

- *We say $\mathbb{S}_{\epsilon,t}^{\mathrm{sin}}(\mathsf{P}_0) \gg \mathbb{S}_{\epsilon,t}^{\mathrm{sin}}(\mathsf{P}_1)$ (or equivalently, $\mathbb{S}_{\epsilon,t}^{\mathrm{sin}}(\mathsf{P}_1) \ll \mathbb{S}_{\epsilon,t}^{\mathrm{sin}}(\mathsf{P}_0)$), if the following equation holds*

$$\inf \mathbb{S}_{\epsilon,t}^{\mathrm{sin}}(\mathsf{P}_0) \ge \sup \mathbb{S}_{\epsilon,t}^{\mathrm{sin}}(\mathsf{P}_1). \tag{13}$$

Lemma 1 (Amplification). *If there exists some t-adversary $\mathcal{A}_0 = (\mathsf{S}_0,\mathsf{R}_0)$, such that for any positive integer c, and for any $\epsilon \ge \lambda^{-c}$, we have*

$$\Pr_{x \in_R \{0,1\}^n} [x \in \mathbf{G}_x^{\mathsf{S}_0,\mathsf{R}_0}(\ell, \varsigma(\ell,\epsilon)+1-\ell, 0.5+\epsilon, 0.5+\epsilon)] \ge \mu \tag{14}$$

then there exists some $t \cdot \Theta(1/\epsilon)$-adversary $\mathcal{A}_1 = (\mathsf{S}_1,\mathsf{R}_1)$, such that

$$\Pr_{x \in_R \{0,1\}^n} [x \in \mathbf{G}_x^{\mathsf{S}_1,\mathsf{R}_1}(\ell, \varsigma(\ell,\epsilon)+1-\ell, 1-negl(\lambda), 1-negl(\lambda))] \ge \mu \tag{15}$$

where λ is the security parameter and $negl(\cdot)$ denotes some negligible function. (The proof is in our full version [28])

---

[12] The reason behind the definition of $\varsigma(\ell,\sigma)$ (i.e. Eq. 10) is explained with details in our full version of this paper. Informally speaking, some steal algorithm $\mathsf{S}(\ell)$ is able to convey *almost* $\ell+1$ bits message to R algorithm, since $|\{0,1\}^{\le \ell}| \approx |\{0,1\}^{\ell+1}|$. When the error bound $\epsilon \ge 2^{-(\ell-1)}$, we do not care the difference between such "almost" $\ell+1$ bits message and actual $\ell+1$ bits message.

Definition 4 (Strong Steal-Entropy Rate in Input). *Let* $P : \{0,1\}^n \rightarrow \{0,1\}^m$ *be a deterministic single-input algorithm. We define the infimum and supremum of steal-entropy rate of algorithm* P *as*

$$\mu^{\perp} \stackrel{def}{=} \frac{\inf \mathbb{S}^{sin}_{\epsilon,t}(P)}{n}; \qquad \mu^{\top} \stackrel{def}{=} \frac{\sup \mathbb{S}^{sin}_{\epsilon,t}(P)}{n} \qquad (16)$$

(Note that this is a counterpart notion of "entropy rate" or "leakage rate".)

Theorem 2 (Separation between Steal-Entropy and Strong Steal-Entropy). *There exists a constant* $c > 0$, *such that for any positive integer* N, *we can construct an algorithm* P, *such that* $\sup \mathbb{S}^{sin}_{\epsilon,t}(P) \leq c$ *and* $\inf \mathbb{S}^{in}_{\epsilon,t}(P) \geq N$. *(Proof is in our full version [28])*

5   Our Proposed Encryption (or Encoding) Scheme

We will describe our proposed encryption scheme in two steps following a modular design.

5.1   Our Steal-Resilient Encryption (or Encoding) Scheme

Definition 5 (Steal-Resilient Encryption/Encoding). *Let* $\Phi =$ (KeyGen, Encrypt, Decrypt) *be a* length-preserving *encryption scheme. Let algorithm* SUFFIX$_\Phi$ *be defined as below*

$$\text{SUFFIX}_\Phi(k; x) = C_1, \text{ where } k := \text{KeyGen}(1^\lambda)$$
$$\text{and } C_0 \| C_1 := \text{Encrypt}(k; x) \text{ and } |C_1| = \tau |C_0|. \qquad (17)$$

Let n *denote the length of plaintext. We say* Φ *is a* $\delta(n)$-*steal-resilient encryption scheme with split-factor* τ, *if the algorithm* SUFFIX$_\Phi$ *has infimum of strong steal-entropy rate* $\mu^{\perp} = \frac{\inf \mathbb{S}^{sin}_{\epsilon,t}(\text{SUFFIX}_\Phi)}{n} \geq \delta(n)$, *where* $\delta(n) \in [0,1]$ *with 1 meaning the best and 0 meaning the worst,* $t = O(poly(\lambda))$, *and* $\epsilon \geq \lambda^{-c}$ *for some positive integer* c.

We remark that, under our definition, most existing encryption schemes (including any existing block cipher under any existing mode of operation, and All-or-Nothing Transform by Rivest [25], and Leakage resilient encryption[13] [1,2,14,17,23,27,31]) are poorly $\delta(n)$-steal resilient encryption with $\delta(n) = 1/\Theta(n)$ approaching to zero when n approaches to infinity.

We found that the linear transformation with Vandermonde matrix is a good steal-resilient encryption scheme. Let ρ be some positive integer (e.g. 8 or 16 or 32) and $GF(2^\rho)$ be a finite field with order 2^ρ.

---

[13] We remark that some of these cited leakage resilient cryptography works actually propose leakage resilient pseudorandom generator/functions, instead of an encryption scheme. These pseudorandom generator/functions can be converted into encryption scheme using classical methods. These resulting encryption schemes will be a poor steal-resilient encryption.

We construct an encryption scheme $\Phi_0 = (\mathsf{KeyGen}, \mathsf{Encrypt}, \mathsf{Decrypt})$ as below.

$\Phi_0.\mathsf{KeyGen}(1^\lambda) \to \mathbf{M}$

1. Randomly choose a $\zeta \cdot (1+\tau)$ by $\zeta \cdot (1+\tau)$ Vandermonde matrix[14], and denote its transpose matrix as $\mathbf{M} = (M_{i,j})_{i,j \in [1,\zeta \cdot (1+\tau)]}$, where $M_{i,j} = \alpha_j^i \in GF(2^\rho) \setminus \{0\}$. The inverse of matrix $\mathbf{M}$ exists and is denoted as $\mathbf{M}^{-1}$.
2. Output $\mathbf{M}$.

$\Phi_0.\mathsf{Encrypt}(\mathbf{M}; \boldsymbol{x})$, where $\mathbf{M}$ is a $\zeta \cdot (1+\tau)$ by $\zeta \cdot (1+\tau)$ matrix and $\boldsymbol{x} \in GF(2^\rho)^{\zeta \cdot (1+\tau)}$ is a row vector of dimension $\zeta \cdot (1+\tau)$ (equivalently, 1 by $\zeta \cdot (1+\tau)$ matrix)

1. Compute product $\boldsymbol{y} := \boldsymbol{x} \times \mathbf{M}^{-1}$ of two matrix $\boldsymbol{x}$ and $\mathbf{M}^{-1}$.
2. Treat $\boldsymbol{y}$ as a bit string with length $(1+\tau)\rho\zeta$ bits, which is the concatenation of $\zeta(1+\tau)$ number of ordered ρ-bits finite field elements.
3. Let $\boldsymbol{y}_0$ be the prefix of $\boldsymbol{y}$ with length equal to $\rho\zeta$ bits.
4. Let $\boldsymbol{y}_1$ be the suffix of $\boldsymbol{y}$ with length equal to $\tau\rho\zeta$ bits.
5. Output $(\boldsymbol{y}_0, \boldsymbol{y}_1)$.

$\Phi_0.\mathsf{Decrypt}(\mathbf{M}; \boldsymbol{y}_0, \boldsymbol{y}_1)$

1. Let $\boldsymbol{y}$ be the concatenation of $\boldsymbol{y}_0$ and $\boldsymbol{y}_1$.
2. Parse bit-string $\boldsymbol{y}$ as a row vector of dimension $\zeta(1+\tau)$ where each vector element is from $GF(2^\rho)$.
3. Compute matrix product $\boldsymbol{x} := \boldsymbol{y} \times \mathbf{M}$.
4. Output $\boldsymbol{x}$.

We remark that, any linear transformation with an invertible matrix could constitute an information dispersal algorithm [24], but is unlikely a steal-resilient encryption.

Our experiments in a Macbook Pro Laptop with Intel i5 CPU (purchased in 2014) show that the encryption or decryption can be done in 0.037 s (about 21 megabytes per second) with a single CPU core when dimension of $\mathbf{M}$ is 12800 and $\rho = 16, \tau = 31$; and in 0.149 s when dimension is 25600 and $\rho = 16, \tau = 63$.

Theorem 3. *Let $\boldsymbol{x} := \boldsymbol{y} \times \mathbf{M}$ be as stated in the above scheme. Then $\boldsymbol{x}$ follows (ζ, ρ)-Blockwise-Uniform distribution, as defined in Definition 1 on page 9. More precisely, parse $\boldsymbol{x}$ as a sequence of elements $(x_1, x_2, \cdots, x_i, \cdots, x_{\zeta(1+\tau)})$ with each element $x_i \in GF(2^\rho)$. If the last $\tau \cdot \zeta$ elements of $\boldsymbol{y}$ is given and fixed, and the first ζ elements of $\boldsymbol{y}$ uniformly distributes over $\{0,1\}^{\rho\zeta}$, then any tuple of ζ elements $(\cdots, x_{i_j}, \cdots)_{j \in [1,\zeta]}$, with distinct indices i_j's, will have exactly $\rho \cdot \zeta$ bits Shannon-Entropy (i.e. the Shannon-Entropy rate is 1). Proof is in full version [28].*

---

[14] The matrix row/column index starts with either zero or one, makes no essential difference to the property of Vandermonde matrix.

Corollary 4. *The proposed scheme Φ_0 is a $\delta(n)$-steal-resilient encryption, with $\delta(n) = \frac{1}{\rho(\tau+1)}$ independent on plaintext length $n = \rho\zeta(1 + \tau)$, and $\inf \mathbb{S}^{\sin}_{\epsilon,t}(\text{SUFFIX}_{\Phi_0}) \geq \zeta$. We remark that both ρ and τ are system parameters independent on plaintext length n. (Proof is in our full version [28])*

We observe that, in the proof of Theorem 3, we only require the first ζ rows of matrix $\mathbf{M}$ satisfy the special Vandermonde matrix property. Therefore, we could simply tweak the rest rows of matrix $\mathbf{M}$, in order to speed up the decryption performance.

Corollary 5. *In algorithm Φ_0.KeyGen, change the last $\tau\zeta$ rows of matrix $\mathbf{M}$ to a sparse matrix, such that $\mathbf{M}$ is still invertible. Then the resulting variant version of Φ_0 is still $\delta(n)$-steal-resilient encryption, with $\delta(n) = \frac{1}{\rho(\tau+1)}$, where $n = \rho\zeta(1+\tau)$.*

The above Corollary 5 actually separates our notion from secret-sharing scheme: After the tweak in the above corollary, the resulting scheme is no longer a secret sharing scheme.

5.2 Combine Steal-Resilient Encryption and Semantic Secure Encryption

We wish to combine both of the advantage of Steal-Resilient Encryption in leakage setting, and the advantage of semantic secure encryption in standard adaptive chosen message/plaintext attack setting (CCA2/CPA2).

Let Φ_0 be the steal-resilient encryption scheme defined above. Let Φ_1 be a given semantic-secure encryption scheme (precisely, CTR mode of a semantic secure block cipher). Eventually, our encryption scheme Φ_2 is defined as below

- Φ_2.KeyGen(1^λ) $\leftarrow (k, k_0, k_1)$:
  1. Compute key $\mathbf{M} \leftarrow \Phi_0$.KeyGen($1^\lambda$).
  2. Compute key $k \leftarrow \Phi_1$.KeyGen(1^λ).
  3. Output $(k, \mathbf{M})$.
- Φ_2.Encrypt($k, \mathbf{M}; \mathsf{Msg}$) $\rightarrow (\mathsf{C}_0, \mathsf{C}_1)$
  1. Encrypt plaintext Msg using semantic secure encryption to obtain ciphertext $\mathsf{Ctx} \leftarrow \Phi_1$.Encrypt($k; \mathsf{Msg}$).
  2. Split the ciphertext Ctx into two shares using steal-resilient encryption $(C_0, C_1) \leftarrow \Phi_0$.Encrypt($\mathbf{M}; \mathsf{Ctx}$).
  3. Output (C_0, C_1).
- Φ_2.Dec($k, \mathbf{M}; C_0, C_1$)
  1. Merge the two shares C_0 and C_1 as ciphertext $\mathsf{Ctx} \leftarrow \Phi_0$.Decrypt($\mathbf{M}; C_0, C_1$).
  2. Decrypt Ctx as $\mathsf{Msg} \leftarrow \Phi_1$.Decrypt($k; \mathsf{Ctx}$).
  3. Output Msg.

We remark that, in our proposed scheme, for large input size, Φ_1 can run in CTR mode and Φ_0 can run over every $\rho\zeta(1+\tau)$-bit segment in ciphertext of Φ_1 independently.

Theorem 6. *Let Φ_2 be the proposed encryption scheme by combining a steal-resilient encryption Φ_0 and a semantic secure encryption Φ_1. Then Φ_2 is semantic-secure in standard model, and is $\delta(n)$-steal-resilient encryption with split-factor τ in our leakage-model, where $1/\delta(n) = \rho(\tau + 1) + O(1)$. (Proof is given in our full version [28]).*

6   Conclusion

In this work, we proposed a new and strong leakage setting, a novel notion of computational entropy, and a construction to achieve higher security against strong leakage. We separated our new notion from several relevant existing concepts, including Yao-Entropy, Hill-Entropy, All-or-Nothing Transform, Exposure Resilient Function. Unlike most of previous leakage resilient cryptography works which focused on defeating side-channel attacks, we opened a new direction to study how to defend against backdoor (or Trojan horse) and covert channel attacks.

Acknowledgment. The first author is supported by the National Research Foundation, Prime Minister's Office, Singapore under its Corporate Laboratory@University Scheme, National University of Singapore, and Singapore Telecommunications Ltd. The second author is supported by the National Research Foundation (NRF), Prime Minister's Office, Singapore, under its National Cybersecurity R&D Programme (Award No. NRF2014NCR-NCR001-31) and administered by the National Cybersecurity R&D Directorate.

References

1. Abdalla, M., Belaïd, S., Fouque, P.A.: Leakage-resilient symmetric encryption via re-keying. In: Proceedings of the 15th International Conference on Cryptographic Hardware and Embedded Systems, CHES 2013, pp. 471–488 (2013)
2. Alwen, J., Dodis, Y., Wichs, D.: Leakage-resilient public-key cryptography in the bounded-retrieval model. In: Halevi, S. (ed.) CRYPTO 2009. LNCS, vol. 5677, pp. 36–54. Springer, Heidelberg (2009). https://doi.org/10.1007/978-3-642-03356-8_3
3. Alwen, J., Dodis, Y., Wichs, D.: Survey: leakage resilience and the bounded retrieval model. In: Proceedings of the 4th International Conference on Information Theoretic Security, ICITS 2009, pp. 1–18 (2010)
4. Kolmogorov, A.N.: On tables of random numbers. Theor. Comput. Sci. **207**, 387–395 (1998)
5. Barak, B., et al.: On the (im)possibility of obfuscating programs. J. ACM **59**(2), 6:1–6:48 (2012)
6. Barwell, G., Martin, D.P., Oswald, E., Stam, M.: Authenticated encryption in the face of protocol and side channel leakage. Cryptology ePrint Archive, Report 2017/068 (2017). https://eprint.iacr.org/2017/068
7. Barwell, G., Martin, D.P., Oswald, E., Stam, M.: Authenticated encryption in the face of protocol and side channel leakage. In: Takagi, T., Peyrin, T. (eds.) ASIACRYPT 2017. LNCS, vol. 10624, pp. 693–723. Springer, Cham (2017). https://doi.org/10.1007/978-3-319-70694-8_24

8. Ben Othman, J., Mokdad, L.: Enhancing data security in ad hoc networks based on multipath routing. J. Parallel Distrib. Comput. **70**, 309–316 (2010)
9. Bronchain, O., Dassy, L., Faust, S., Standaert, F.X.: Implementing Trojan-resilient hardware from (mostly) untrusted components designed by colluding manufacturers. In: Proceedings of the 2018 Workshop on Attacks and Solutions in Hardware Security, ASHES 2018, pp. 1–10. ACM, New York (2018). https://doi.org/10.1145/3266444.3266447
10. Canetti, R., Dodis, Y., Halevi, S., Kushilevitz, E., Sahai, A.: Exposure-resilient functions and all-or-nothing transforms. In: Preneel, B. (ed.) EUROCRYPT 2000. LNCS, vol. 1807, pp. 453–469. Springer, Heidelberg (2000). https://doi.org/10.1007/3-540-45539-6_33
11. Daemen, J., Rijmen, V.: The Design of Rijndael: AES - The Advanced Encryption Standard (2002)
12. Apon, D., Huang, Y., Katz, J., Malozemoff, A.J.: Implementing cryptographic program obfuscation. Cryptology ePrint Archive, Report 2014/779 (2014). https://eprint.iacr.org/2014/779
13. Di Crescenzo, G., Lipton, R., Walfish, S.: Perfectly secure password protocols in the bounded retrieval model. In: Halevi, S., Rabin, T. (eds.) TCC 2006. LNCS, vol. 3876, pp. 225–244. Springer, Heidelberg (2006). https://doi.org/10.1007/11681878_12
14. Dodis, Y., Haralambiev, K., López-Alt, A., Wichs, D.: Efficient public-key cryptography in the presence of key leakage. In: Abe, M. (ed.) ASIACRYPT 2010. LNCS, vol. 6477, pp. 613–631. Springer, Heidelberg (2010). https://doi.org/10.1007/978-3-642-17373-8_35
15. Dziembowski, S.: Intrusion-resilience via the bounded-storage model. In: Halevi, S., Rabin, T. (eds.) TCC 2006. LNCS, vol. 3876, pp. 207–224. Springer, Heidelberg (2006). https://doi.org/10.1007/11681878_11
16. Dziembowski, S., Faust, S., Standaert, F.X.: Private circuits III: hardware Trojan-resilience via testing amplification. In: Proceedings of the 2016 ACM SIGSAC Conference on Computer and Communications Security, CCS 2016, pp. 142–153. ACM, New York (2016). https://doi.org/10.1145/2976749.2978419
17. Dziembowski, S., Pietrzak, K.: Leakage-resilient cryptography. In: Proceedings of the 2008 49th Annual IEEE Symposium on Foundations of Computer Science, FOCS 2008, pp. 293–302, IEEE Computer Society, Washington, DC, USA (2008). https://doi.org/10.1109/FOCS.2008.56
18. Goldwasser, S., Kalai, Y.T.: On the impossibility of obfuscation with auxiliary input. In: Proceedings of the 46th Annual IEEE Symposium on Foundations of Computer Science, FOCS 2005, pp. 553–562 (2005)
19. HÅsstad, J., Impagliazzo, R., Levin, L.A., Luby, M.: A pseudorandom generator from any one-way function. SIAM J. Comput. **28**(4), 1364–1396 (1999)
20. Krawczyk, H.: Secret sharing made short. In: Stinson, D.R. (ed.) CRYPTO 1993. LNCS, vol. 773, pp. 136–146. Springer, Heidelberg (1994). https://doi.org/10.1007/3-540-48329-2_12
21. McEliece, R.J., Sarwate, D.V.: On sharing secrets and Reed-Solomon codes. Commun. ACM **24**(9), 583–584 (1981)
22. Micali, S., Reyzin, L.: Physically observable cryptography. In: Naor, M. (ed.) TCC 2004. LNCS, vol. 2951, pp. 278–296. Springer, Heidelberg (2004). https://doi.org/10.1007/978-3-540-24638-1_16

23. Pereira, O., Standaert, F.X., Vivek, S.: Leakage-resilient authentication and encryption from symmetric cryptographic primitives. In: Proceedings of the 22nd ACM SIGSAC Conference on Computer and Communications Security, CCS 2015, pp. 96–108 (2015)

24. Rabin, M.O.: Efficient dispersal of information for security, load balancing, and fault tolerance. J. ACM **36**(2), 335–348 (1989). https://doi.org/10.1145/62044.62050

25. Rivest, R.L.: All-or-nothing encryption and the package transform. In: Proceedings of the 4th International Workshop on Fast Software Encryption, FSE 1997, pp. 210–218 (1997)

26. Shamir, A.: How to share a secret. Commun. ACM **22**(11), 612–613 (1979)

27. Standaert, F.-X., Pereira, O., Yu, Y.: Leakage-resilient symmetric cryptography under empirically verifiable assumptions. In: Canetti, R., Garay, J.A. (eds.) CRYPTO 2013. LNCS, vol. 8042, pp. 335–352. Springer, Heidelberg (2013). https://doi.org/10.1007/978-3-642-40041-4_19

28. Xu, J., Zhou, J.: Strong leakage resilient encryption by hiding partial ciphertext. Cryptology ePrint Archive, Report 2018/846 (2018). https://eprint.iacr.org/2018/846

29. Xu, J., Zhou, J.: Virtually isolated network: a hybrid network to achieve high level security. In: Data and Applications Security and Privacy XXXII, DBSec 2018, pp. 299–311 (2018)

30. Yao, A.C.C.: Theory and applications of trapdoor functions. In: Proceedings of 23rd Annual Symposium on Foundations of Computer Science, EUROCRYPT 2007, pp. 80–91 (1982)

31. Yu, Y., Standaert, F.X., Pereira, O., Yung, M.: Practical leakage-resilient pseudorandom generators. In: Proceedings of the 17th ACM Conference on Computer and Communications Security, CCS 2010, pp. 141–151. ACM, New York (2010). https://doi.org/10.1145/1866307.1866324

Author Index